Foreword

by The Lord Rogers of Riverside

New knowledge leads to technological breakthrough if changes are followed by times of consolidation and refinement. For example, man's ability to navigate at sea was revolutionised by the invention of the chronometer, and his ability to express himself graphically by the development of the rules of perspective.

The designer of today is faced with developments that allow him to express his ideas in ways undreamed of just a few years ago. Whether for the design of an aircraft, an automobile, or a building, computer-based 3D modelling offers the designer a powerful, new visually realistic and analytically powerful medium with which to create and express his ideas accurately. This, I believe, will be a key factor leading eventually to the design of buildings of greater quality for both users and for the community at large.

The challenge before us now is to seize this advance and its consolidation and refinement with vision and enthusiasm. It will be a challenge to all, in particular to those who are training the next generation of architects and other designers, to harness effectively the new levels and freedom that 3D CAD modelling offers.

The authors of this study identify clearly the key issues surrounding project modelling in construction and demonstrate by means of case studies the progress achieved.

Post-industrial information technology is revolutionising the way we think and will change our work radically in the next decades.

Richard Rogers

Pr
in
...se

Authors

Norm

Richard Barlow (project manager)

Naomi Garnett

Edward Finch

Robert Newcombe

Steering group

Gerry Beers (chairman), *Taylor Woodrow*
Nigel Graham, *Whitbread*
Dr Nicholas Heaf, *Coca Cola*
Dr Neil Jarrett, *DoE*

David Smith, *Taylor Woodrow*
Nick Terry, *BDP*
Raymond Turner, *BAA*
Nigel Woolcock, *Prudential*

TAYLOR WOODROW   Sponsored by Taylor Woodrow Management Ltd

Published by Thomas Telford Publishing, Thomas Telford Services Ltd, 1 Heron Quay, London E14 4JD

First published 1997

Distributors for Thomas Telford books are
USA: American Society of Civil Engineers, Publications Sales Department,
345 East 47th Street, New York, NY 10017-2398
Japan: Maruzen Co. Ltd, Book Department,
3-10 Nihonbashi 2-chome, Chuo-ku, Tokyo 103
Australia: DA Books and Journals, 648 Whitehorse Road, Mitcham 3132, Victoria

LEGO® is a registered trademark of LEGO A/S, Denmark
Windows™ is a trademark of the Microsoft Corporation

A catalogue record for this book is available from the British Library

ISBN: 0 7277 2581 5

Designed and typeset by Rob Norridge

Printed in Great Britain by Spottiswoode Ballantyne Printers, Colchester

Preface

As the globalisation of trade increases, multinational companies find themselves accelerating expansion of their business infrastructure and, at the same time, needing higher value added from each investment. Both of these objectives demand a significant reduction in the time taken to make an investment decision, but, with the future of the business at risk, no reduction in the quality of the business case information is acceptable.

Leading international companies are using project modelling/prototyping to test investment business cases internally with a view to shortening the decision making process from months to weeks, and in some cases days, at the same time increasing the completeness, quality and consistency of the information presented. This publication describes and illustrates the thinking, technology and process reengineering behind such performance improvements. What is described takes risk out of the 'one-off' project, because for the first time it is possible to develop an affordable realistic 3D prototype electronically.

For clients, the use of electronic prototypes, visualisation and the harnessing of user knowledge, will result in more certainty in terms of cost, time and quality targets, and the gradual improvement in value for money, operating costs and buildings that incorporate lessons learned elsewhere. The better preplanning, in all its forms, will slowly achieve the improvements, not from cutting fees, scope, specification or from robbing players of a reasonable profit, but from performance improvements that gradually eat into the current 'built in' inefficiencies that have evolved with the traditional fragmented system.

This publication grew from ideas expressed some time ago by two people. By Norman Fisher in his inaugural lecture as the BAA Professor of Construction Project Management. His lecture was entitled *Construction as a Manufacturing Process?*, and presented at The University of Reading on May 18th 1993. Also by Gerry Beers, a director of Taylor Woodrow, in a speech at The Institute of Directors on 29th October 1993, entitled *Project Modelling in Construction*.

A number of others have also played key roles in the development and testing of the technology and the new ways of working. They work in client and contractor organisations; university research groups, such as ACT-Reading; or specialist IT companies. They include the authors and the sponsors of this publication and members of its steering group.

Since 1993, the ideas have been developed and tested on designs, projects and components, with a variety of different organisations. Some of the developments and the testing are described in the text of this publication.

Acknowledgements

The authors wish to acknowledge with gratitude the help and support of many people who have given encouragement to this publication. In particular, they are grateful to the sponsors of the study and to the steering group for their invaluable help and advice. Thanks are also due to the designers and contractors who so willingly provided material for the case studies, to several major industrial organisations and government funding agencies that supported the research and, finally, to numerous colleagues from industry and academia worldwide who helped us develop and prove the ideas.

Contents

Chapter 1

Introduction

t was Benjamin Disraeli, the first Earl of Beaconsfield, much influenced by the effects on society of the Industrial Revolution, who said

Change is inevitable in a progressive country. Change is constant.

He could have been talking just as easily about an industry, not a country.

A number of major changes are taking place today similar in scope to those which took place in the Industrial Revolution of 1770–1850. We are in the midst of another revolution, this time an electronic revolution. During the Industrial Revolution man learned to extend his arm and his muscle, today he is learning to extend his brain – both intellectual skills and knowledge manipulation. Then, the Industrial Revolution profoundly affected blue collar workers, but today's revolution also profoundly affects knowledge workers – managers, accountants, engineers and designers. However, the resistance to change, the protection of the old or traditional ways, is just as great today as it was then. What Britain, the home of the Industrial Revolution, did then was to be *Great*; men and women of vision, courage and a desire to create wealth, overwhelmingly embraced the new ways, the new technology and the new processes. Not all the results were good, not everyone benefited at first, but the acceptance of the challenge caused by the new ways led to a better age for almost all: better medicine, better food, cheaper manufactured goods, better travel and better homes. Today, we have a not dissimilar challenge, one that will affect the well-being of future generations.

At the core of this publication is a debate on how (starting now) we can use and transfer knowledge and skills in the future. It looks at the competitive advantage that use of the proven technology, described and currently available, will bring. The publication includes a large number of case studies of the use of the technology, not pie-in-the-sky ideas, but examples of advanced IT and processes delivering now in commercially tough situations.

Johann Gutenberg in the 15th century revolutionised the way we store and transfer information by the invention of movable printing types. Today, advanced CAD, object oriented technology and virtual reality linked to the likes of the Internet are doing the same thing. Design-to-manufacture industries such as the car industry and other design-to-order industries such as the petrochemical industry have been quick to use these technologies to revolutionise the way that they do business successfully. A good example is how Jaguar Cars have brought together object oriented technology and virtual reality in the design and development of their new XK8 model.

The development and successful deployment of these technologies have allowed the users to develop an increasingly market and user driven approach to the design and manufacture of their products. The need for such an approach in construction was highlighted in 1993 by a major study entitled *The UK Construction Challenge,* published by the London developers Lynton plc.

With the successful deployment of the technologies and process improvements outlined in this publication, the construction industry will be in a position to more than meet the challenge to show innovation in delivering the improved value that clients and the nation are demanding; at the same time improving the quality of the built environment through better and cheaper design and construction. The next generation have a right to expect this and to benefit from it as we have from the Industrial Revolution.

Object oriented modelling

Imagine a refurbishment project to be carried out in the space of three weeks on a building of national cultural significance; a project which needed to be completed in time for the appearance of, for example, Pavarotti, the London Symphony Orchestra, royal guests and dignitaries. Just such a scenario arose for the refurbishment of the Royal Albert Hall balcony seating. How was this to be achieved? Conventional construction and project management methods could not provide the kind of certainty required to hit the target date. To achieve success a radical solution was required; one that would allow the programme planners to envision the project down to the most precise detail before activity began on site, with no room for surprises, no room for uncertainty. Taylor Woodrow's answer as the construction manager was to use project modelling; an approach which would entail major changes in the way that the prototyping and consultation process would be undertaken. An approach which would enable stakeholders to ask searching questions about what they wanted, what they could do, and how they could do it.

In August 1995 tenders were submitted for the refurbishment programme of the Hall's balcony level seating which was to be upgraded to dress circle standard. The visibility of the stage from the balcony seats needed improvement, because of obstructions from handrails and people immediately in front. The answer was to increase the inclination of the seating tier to improve general seat visibility. Using an object oriented model it was possible to reproduce the visibility of the stage from any of the 1500 seating positions on the balcony. In

the theatre world exits and entrances are known as vomitories and these also obstructed sight-lines because of rails and banisters. Using the model designers were able to redesign these and reduce the number of entrances and exits from thirteen to five.

So, object oriented modelling helped to produce a convincing solution – but did it make it any easier to build? Could it be built on time? The model was used with contractors and the client at the detail design stage to address buildability issues. Managers representing the consultant, the building contractor, and the seating contractor participated in workshops in which a large computer screen was used to display the sequencing of work on a daily basis. Working in cramped and confined conditions would inevitably lead to delays, so the execution of each of the tasks had to be thought out to the smallest detail, with the overlap of seat removals, tier replacement and new seat installation. Only through object oriented modelling could this kind of programming issue have been examined in such detail before any work began on site.

So, did it save the client any money? Well, one of the outcomes from the careful remodelling of seating positions was the realisation that a further five seats could be accommodated at the balcony level. Not much, you might say. But if you consider the takings which will accumulate over the course of a season, things start to look interesting.

Perhaps the question we are left with for a project of this kind is not whether object oriented modelling was needed, but whether the project could have gone ahead without it.

Information on the balcony seating refurbishment was kindly provided by Ian Blackburn, Property Director at the Royal Albert Hall.

Chapter 2

What is object oriented modelling?

Everyone has a different world view. Depending on their expertise and interests, each individual concentrates on aspects that they see as important and ignores those that they see as irrelevant.

This is never more apparent than when consultants and contractors come together to design a building. The problems caused when different participants' views of the same task differ are well recognised; there is enough truth in these pictures of the construction of a swing to raise a smile from anyone with experience of the construction process. But, for many, communication failures that prevent the delivery of successful projects are so common that the joke is wearing thin.

What was included in the client's requirements ?

What was included in the contractor's proposals

The architect's vision

What was included in the budget;

Designing and constructing a modern building requires an input from a diverse range of consultants, each with a particular expertise. In order to progress their part of the project, each player needs to concentrate on a subset of all the information needed to build the building. Difficulties occur where these subsets overlap. Unless these intersections are carefully managed, anarchy results.

There are few who would argue that existing project processes hold all the answers to the industry's communication problems. The latest developments in object oriented modelling provide a major step forward. By simplifying the process, object oriented modelling allows a computer to take over the management of the intersection between each project participant's work. To understand the concept of object oriented modelling, it is best to begin by considering a simple example of the design process.

A client wants to build a simple office wall. There is a list of requirements for this wall

• it must look good and divide the existing office into useful spaces

• it must stand up and carry load transferred from the slab above

- it has to resist sound transmission through it and hold conduits for light switches and power sockets

- it must not cost more than the allocated budget.

The client knows that all these problems cannot be solved by one person so a construction team, made up of specialists – architect, structural engineer, services engineer, cost consultant and construction manager – is employed to design the wall. Each team member is chosen because they have a particular skill needed to ensure that the final design meets all the client's requirements and hence they will concentrate on that aspect of the design.

The architect, for example, deals with the aesthetics and space management, and the structural engineer tackles stability and strength related issues. The services engineer takes on the section of the brief relating to acoustics and power supply, while the cost consultant keeps an eye on the budget as fellow team members develop their ideas. At the same time the construction

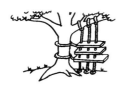

What was included in the engineer's final detail

What was included in the QS's measurement

How the project was finally constructed

What the client actually wanted

manager is considering how the wall will be built. Although each consultant has a specialised skill to apply, they need information generated by other consultants to perform their duties. For example, the structural engineer will need to know the dimensions chosen by the architect for the wall. The services engineer will want to know the materials chosen by architect and structural engineer to calculate the acoustic performance. If the services engineer decides to enclose cable ducts in the wall the structural engineer will want to know where so that the effect on the wall's strength can be calculated. The cost consultant will want to know changes to materials so that the estimated cost can be adjusted and checked against the budget. And the construction manager will want to know what the proposed materials are so that a construction time can be estimated and a programme compiled which may in turn impact on the budget.

It is managing this complex flow of information between each team member, ensuring that each member is up to date on the latest revision, that causes problems. Electronic *objects* can provide the solution.

An object is an electronic representation of an element of a building, such as a door, wall or window. Attached to each object is a series of electronic tags, each containing one piece of information, or attribute, of the object. Attributes for a typical building object would include geometry, position, orientation, material, colour, cost, serviceability, and relationship to other objects – in essence, any property which a user or the system needs to know about.

These objects can be assembled like children's interlocking building blocks, such as Lego® System bricks, to form a bigger model. For instance, objects representing doors, walls, windows, roofs and floors could be assembled into a room. For this to work the objects must have all the information, in the form of attributes, needed to use them in a bigger assembly. There must also be mechanisms in place for information to be passed between objects in an organised way. There are several methods of relating components which demonstrate the power of object orientation. The simplest link is to create an association between objects. Two associated objects are grouped together and can be moved around as a single object. This is the relationship which will be familiar to most users of typical Windows™ programs.

Using interactive objects as building blocks is the key to creating effective computer models. By adopting an object oriented approach, building components, activities and information can be created in a way which allows them to be used in a variety of different situations. In object oriented design, each object is composed of a number of other objects, descending to the required level of detail. An object representing a room can be made up of objects representing doors, windows, walls, roof and floor. To make life easier, by arranging objects in a hierarchy, attributes at lower levels can pick up their values by inheritance. Thus inheritance means that objects lower down the hierarchy (children) inherit attributes from objects that are higher up (parents) without further intervention by the user.

This means that changes can be made globally, at a high level with a minimum of effort, and these changes then ripple through the model. For example, major changes such as altering the number and dimensions of building grids – normally a daunting task for a traditional CAD system – become easier for the user, with the time taken to complete the task simply depending on the computer power available.

In turn, local changes can be made to particular children if they need to depart from the more generally used values. This could happen around service cores, where the normal constraints on, for instance, the structural depth could be made more onerous to accommodate the increased size and complexity of air-conditioning ductwork.

A more powerful object relationship is called polymorphism (from the Greek 'many shapes'). A new object can be defined with a minimum of extra work by building on an existing object. Creators of new objects do not need information on the internal workings of the original ones; they simply define a new object as being a type of 'existing object' and then add any new attributes required.

Once a model of a building is constructed from objects, if someone decides that a new body of knowledge is needed, then new objects and attributes can be added with a minimum of fuss, and without adversely affecting the way the existing attributes behave.

Likewise the design and construction team can apply their own filters to the information, screening out extraneous details and hence reducing it to manageable proportions. But no matter which attributes they choose to work from, because they are contained in the one model they will be consistent with those used by everyone else in the team. The confusion caused by different views of the tree swing is prevented.

Objects as described above are linked to each other within a model, but ideally we would also like a model to be able to link to other systems. For many years the ideal of a fully integrated system has been an unattainable vision. The evolution of systems is neatly described in a diagram entitled *Islands of Automation* by Matti Hannus. This shows the islands emerging from the ocean and is based on the premise that by the year 2000 the new coastline will enclose them all! (More information is available on the web site given in the references at the end of the report.)

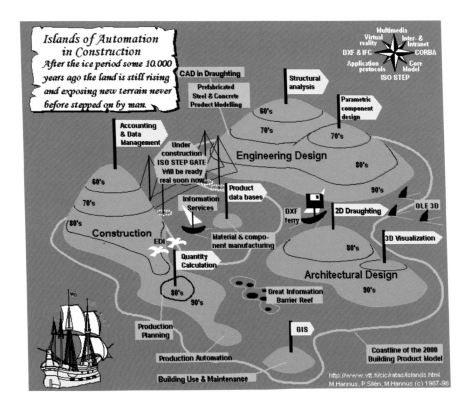

A common confusion exists between object oriented modelling and the systems which use it, and traditional CAD systems which often allow the user to work in three dimensions and produce 3D models and visualisations. To the casual observer the results are often difficult to distinguish, but the key differences lie in the way information is entered and stored. A simple traditional CAD system is project specific and re-using any of the information is difficult if not impossible. Additionally, the idea of associating a whole range of properties with the simple shapes making up the CAD model is beyond the capability of many traditional CAD systems. However, an object oriented modelling system stores information about objects, often in an external database, in a manner which allows re-use. This means that starting a new project does not necessarily mean starting from scratch, but instead from existing libraries of objects. Similarly, a knowledge based system stores not only the objects, but also the rules necessary to assemble them – a potentially enormous time saving when it comes to starting a new project.

The issue of how to share data originating in different organisations is critical. Translation is a solution which has been used extensively in the past, but it inhibits system performance and introduces a continuing need to update the translators with every new version of a software package.

The Standard for the Exchange of Product Model Data (STEP) standard (ISO 10303), is intended to simplify this problem, by agreeing a single universal data standard for each domain. Work has been continuing for more than ten years. Impatient with getting a usable standard for interoperability, there are now commercially backed initiatives which are gaining momentum, and which do not necessarily exclude STEP.

Some of the key issues and objectives being addressed in the move towards interoperability are

- Clients are expecting flexibility and quality in their buildings so they can stand up to the pressures of their current business.

- The design time of products has decreased from years to months and the modern facility, being essentially another product, is expected to keep pace in its method of creation and management of changes.

- It is estimated that up to 30% of the cost of building is lost due to the current processes in the AEC industry. Building owners demand facilities that are cheaper to construct, maintain and operate.

- Today, buildings contain products produced around the globe, and the whole process for creation of new buildings is quickly becoming global.

The intention is to develop a common language, called the industry foundation classes for technology, to improve the communication, productivity, delivery time, cost and quality throughout the design, construction and maintenance life cycle of buildings, for information sharing through the project life cycle and across disciplines and technical applications. There are also significant initiatives supported in the UK by the Department of the Environment's Construction Sponsorship Directorate which address these issues.

Why does the construction industry need object oriented modelling?

bject oriented modelling offers its users the ability to view their particular world as a number of objects which can be related to each other according to defined rules. Each object can have characteristics or attributes of very different natures, from cost to material density, assigned to it. Coupled with modern day computing technology and graphics, object oriented modelling opens up new vistas in not just 3D but also 4D if the effects of time on the model are also included.

This technology is being applied in areas – from the design of cars and aeroplanes to business process reengineering – where business processes are depicted as objects. The ability to use object oriented modelling at all stages of design from inception to finished product has proved particularly attractive in industries where cost, time and getting it right first time are paramount.

But why does the construction industry need this technology? Why do we need to change the way in which we design and erect our buildings? And what will it offer that is currently not available? To answer these questions it is necessary to look at two levels within the industry; firstly, at the strategic, or macro level across the industry as a whole to identify why we need to change, and secondly, at project or construction business level, i.e. the micro level, to discover what object oriented modelling will improve in terms of creativity, communication, certainty and coordination – the critical success factors of the construction process.

Having established that there is potential for the use of object oriented modelling in construction, let us consider the nature of that potential and what it will enable. The final point is that no matter how good object oriented modelling is or how high its potential, if the business case cannot be made, what benefit will occur? The business case is identified and assessed on page 11.

The construction industry today – a macro view

The UK construction industry is a vibrant, hard working industry with more than its fair share of talented people. However, a careful analysis of it reveals a paradox. Numerous organisations – architects, surveyors, consulting engineers, contractors and component manufacturers and suppliers – are regularly winning competitions and orders and are operating successfully worldwide. In so doing they are earning the accolade of being world class operators. However, when operating in the UK performance is often left wanting. The UK industry's poor home performance can be measured in a number of ways.

In the report *Strategies for the European Construction Sector* (1994) by WS Atkins International Ltd for the European Commission, construction industries in a group of countries including all EU members, the US and Japan were ranked in terms of purchasing power parity. UK construction as a finished product was second worst, only slightly better than Greece. Put simply, UK construction when an average is taken over a full economic cycle, is a high cost–low wage industry.

It is widely accepted that the fragmented nature of the UK construction industry, perhaps more so than any other EU member, is a major factor in this low home performance. Poor communication and coordination results from the fragmented way of working that has evolved in the UK. The industry is based around a multitude of professions supported by their appropriate 'protecting' institutions. Each profession contributes their 'bit' to the process in a territorial manner, but there is insufficient management of the interfaces. This has led to low levels of productivity in design and on site, and to an adversarial culture, which in turn has resulted in cost overruns, poor quality and widespread client dissatisfaction.

Emerging engines for change
There have been many initiatives since World War 2 attempting to improve productivity, data coordination, communications and quality. Perhaps the most significant during the late 1980s and early 1990s has seen clients and other capital providers demand improvements in performance, measured in terms of better value for money, higher quality and lower running costs.

While different clients have acted in different ways, their basic demands have been similar. Their business requirements have, however, been different. There are three principal groups of clients

(a) Property developers who are keen to sell on a building and/or demand short-term returns on their investments. Active first in the mid 1980s the concerns of this group have centred around speed of construction and design issues that maximised property sale or rental value. The objective, to optimise developer profit, is often incompatible with the business of other groups of clients.

(b) Regular clients of the construction industry who are, in effect, adding capacity as a solution to an identified business opportunity. Such clients usually hold a large property portfolio and can exert considerable influence over parties to the construction process. Such groups include utilities, airport operators and retail chains. They seek continuous improvement in the value for money that they receive from their capital expenditure. They are concerned with how a building performs during its full economic life.

(c) Occasional clients of the industry who, although accounting for a significant part of construction activity, are very much at the mercy of the traditional

The business case for object based 3D modelling

The view of one large regular client of the industry

A client will use this technology to help them gain an accurate understanding of what they are getting in terms of design-for-money. That is better design, better use of space, better quality and more effective use of design time.

The benefits are achieved firstly by graphic visualisation for walk-through to help non-technical stakeholders, perhaps within a client organisation, understand the design and contribute their expertise to the process of obtaining better value for money. For example, a signage scheme around a large complex building to which the public have access appeared to be effective on drawings, but in 3D was quite obviously inadequate. The 3D solution that was finally agreed upon turned out to be a very accurate representation of how the signage system finally appeared and worked when the building was constructed.

Secondly, the concept of object oriented electronic prototyping offers potentially greater benefits to a client in particular for the increased use of standard components and pre-construction digital mock-ups. So far, as little as 10–15% of the full benefit of the use of this technology has been achieved. However, the potential benefits of its use will only be achieved if used by a client with a partnering approach, using dedicated teams and by careful management of the supply chain. Further, these benefits will only be achieved if in addition they are driven by a strong client. As one major client put it, if left to the industry *it could take as long as 20 years for the current technology to be effectively and widely used*.

Designers and, in particular, architects are faced with the challenge to be *genuinely* committed to the technology and not just pay lip service.

The key business issues for a client

(a) With a 3D modelling approach effectively used, sufficient value to justify its use can currently be obtained if the extra cost of the technology is not greater than 10% of the total design fees.

(b) Effective use of the technology to a client means that it will

   (i) help a client/stakeholder understand what they are getting for their money in terms of value, in particular use of space
   (ii) force a more streamlined concurrent design process – one (electronic) drawing board.

Other client deliverables expected from the use of the technology include

- the benefits of more complete design and better preplanning
- excellent records of 'what is there' during the project delivery process
- the basis of an as-built model for linking to bar-coding/smart dot technology for maintenance planning/records.

construction process and who look to regular clients to show change leadership. The objectives of this group are close to those of the regular clients group above.

The cost of construction has come under considerable pressure at all stages in the traditional building cycle. International comparisons have shown widely different country costs for ostensibly similar products. Clients attacking their cost base in all areas of their business see accommodation as just another area for improvement. They do not see why they should pay significantly more in one country than another without good reason. Expectations of cost reduction by the use of better business processes, supported by IT advances, have been raised by Sir Michael Latham and others. Further, studies in the US and elsewhere have shown a direct link between better buildings and better company performance, whether offices, car plants, or dealing rooms for merchant banks.

Since the 1960s the way in which contractors organise their work has changed dramatically. The number of parties to a project has increased with the emergence of the specialist contractor. Each specialist is responsible for the design, manufacture and assembly of their specialism. To cope with the increased organisational complexity, project management and construction management have become more popular. This trend is likely to continue as the industry reaps the cost advantage of componentisation, resulting from developments in advanced manufacturing such as digitally controlled assembly lines and mass customisation.

Advances in IT have given an opportunity for construction and design companies to redesign not only their business processes but also the design and construction process itself. This can dramatically reduce the man-hours per unit of design and thus allow more to be offered to a client for lower fees.

If design is split into design-to-manufacture and design-to-order, much can be learned from the experience of other design-to-order industries such as the process, petrochemical and aerospace industries. Certainly, man-hour reductions seem possible at all aspects of the design stage if advanced IT is used and would represent a good starting point for a construction client.

A successful advance in other design-to-order industries that can be transferred to construction is the concept of an electronic or digital prototype. Design-to-order industries can use this technology to develop cost effective prototypes of one-off projects and so reap the benefits of prototyping that are taken for granted in design-to-manufacture industries such as the car industry. The use of this approach would suggest a 'quick win' for construction.

Electronic prototypes of construction projects could provide many benefits. For developing business ideas, at the feasibility and the concept design stages, electronic prototypes can be used to harness knowledge from previous projects and from the company's fund of knowledge of working with similar facilities. In addition, both designers, client staff and potential non-technical users can see a visualisation of a proposed design and make suggestions based on their operational experience.

As was described earlier, object orientation allows prototypes to be built that are more than the coloured boxes that were constructed with traditional CAD (2.5D and 3D). Such prototypes contain the rules and algorithms not only of the constraints of design, but also of facilities planning, linked to the client's business. A realistic architectural design/engineering driven visualisation is then possible using state of the art graphics.

Many branches of the design-to-manufacture industry group and of sections of design-to-order have shown that digital prototypes will provide a number of clear deliverables. A list would include

- better value for money through more complete design, due to the harnessing of design knowledge and operator/user experience

- models that can be presented in the most appropriate form for each stage of the project process: feasibility, concept, coordinated design, componentisation, digital mock-up, construction and hand-over; and the business process: business case development, evaluation, design management, cost management, supply chain management, construction management and commissioning; of the client company. (These can be achieved through an integrated project database. An example of different model representation might be as a process flow model at feasibility, an interlocking block model at concept and a full visualisation at both concept and coordinated design stages.)

- better information management and design coordination

- the incorporation of more standard (roll-over, i.e. used before and understood) components, and fewer components in total

- faster fabrication on site through better dimensional coordination, better connections, better assembly planning and training using digital mock-ups

- better involvement of the component manufacturer, through information coordination and sharing, joint digital trial assemblies, group planning and training.

Exciting, challenging, the way forward – but not a panacea

Aggregate savings achieved elsewhere in advanced manufacturing where a combination of fundamental process reengineering and the effective adoption of suitable advanced technology/processes, would suggest Latham's 30% in real terms is achievable without old-fashioned cost cutting (margins, scope and specifications). However, it should be clear that all this is not a panacea; attitudes to change, in terms of understanding why there is no alternative to adopting the technology, are vital. In addition, such changes raise large questions about education, training and a commitment to continuous improvement. All of these issues will be dealt with later in the publication.

What those among our ancestors who were so successful in the Industrial Revolution did and what was perhaps the key to their success, was to realise that a new way of thinking was fundamental, as was a profoundly different approach to traditional ways of working and relationships. Similarly, this is true with the use of advanced IT. A radically different way of thinking is essential and the successful use of the technology will fundamentally change ways of working and professional relationships.

Construction professionals – a micro view

What generic construction activities does object oriented modelling support?

Creativity
Design consultants argue that the move towards manufacturing will seriously damage the industry by removing the ability to be creative in design. In driving down costs the end results will be similar to those of mass production, i.e. a functional building with little aesthetic appeal. This, they are at pains to point out, is reminiscent of the late 1950s tower block approach for which society in general is now having to pay the price.

If the industry is to adopt a manufacturing approach, it is the recent paradigm of lean manufacturing that the industry should be striving towards. The difference between the new paradigm and the mass production characterised by Henry Ford is perhaps best described as the difference between achieving high performance and high productivity. The key is that the measurement of performance is subjective dependent on the customer requirements for the project. It is the continuous search for innovative ways of both determining and meeting client's requirements that generate repeat business and hence ensure long-term profitability.

If, as this new paradigm demands, quality is a foundation stone then how can high-quality designs be achieved within the appropriate cost that the client wishes to pay? The use of object oriented modelling allows the capture of experience and knowledge from earlier projects to be systematically applied to a new one. In this way mundane repetitive tasks can be taken care of by the technology while the innovation is created by allowing the frequent use of the *what if* scenario. Several examples of this will now be given.

Firstly, a knowledge based system could be used to design a hospital for a trust looking to build a high tech, high quality environment for the future. By developing a knowledge based system incorporating all the rules about how a good hospital should be geometrically laid out, a concept model could be built from a small amount of input data, such as number of beds required of a particular type. Clients could consider whether they have the right mix of beds and understand how changes to the mix will alter the basic concept of what they are trying to achieve. If this concept model could be linked to a cost tool, then not only would the change in geometry be apparent, so would the associated cost change. The technology, therefore, would allow clients to be sure that they have the right mix of beds to suit their budgets and future plans.

A second example takes the above idea a stage further with a visualisation package allowing both client and architect to experiment with different aesthetics. Because the changes can be made within a short amount of time, e.g. hours or even minutes, much greater interaction is possible between client and architect on the way to the optimum concept for the project. Before such IT based tools became available this creativity would have taken weeks, and in reality only one or two options may

have been explored. In this way the use of object oriented modelling allows greater creativity to become part of the design process. It does, however, represent a challenge to the traditional relationship between client and designer.

This process could be extended to include the client's customers. For example, customers could be asked to trial the new building through the use of virtual reality and to report back to the design team what was wrong or 'user unfriendly' about their proposal. This increases the creativity of the design team since it allows real time response to ideas and dislikes. There are further uses for the model that support creativity at a more detailed design level where choice of materials and finishes can be modelled to determine colour schemes and atmosphere.

Communication

The fragmentation of the construction industry has a number of effects on the way in which projects are carried out. A major issue among a project team is that of communication. Different cultural and professional backgrounds all contribute to creating a situation similar to that after the building of the Tower of Babel. Project schedules, however, do not permit the luxury of lengthy *getting to know you* situations. Indeed, the lack of communication often results in the adversarial situation common on a large proportion of construction projects. An international project takes this level of communication into another sphere.

Object oriented modelling and the subsequent visualisation of the concept can reduce the *forming* and *storming* phases that project teams often go through before reaching the relatively calmer waters of *norming* and *performing*. It provides a common baseline for individuals on the project team and can become the means by which teamwork is quickly established.

A key area of communication which can be supported in this way is for first time clients needing to build to support their businesses. The foray into the world of construction is often described as extremely stressful. Frequently, a client will only make the decision to build as a final resort. In deciding upon the most suitable building for their needs, first time clients, without any previous experience of 2D drawings and construction technology, must depend upon their chosen project team members.

Often clients are not happy with the end result. To help resolve this, architects build scale models of the proposed schemes. These are costly and have the disadvantage of having limited subsequent use. First time clients could quickly gain confidence in their design teams by the use of object oriented modelling linked to a visualisation tool. They are able to see the outcome of their discussions with the project teams in a true to life form. Further linking of the model to virtual reality would allow them to experience the buildings they are proposing to buy and have the opportunity to assess whether they will fulfil their needs.

Even for sophisticated clients the opportunity to view the goods at an early point in the design process is worthwhile. It builds confidence within the design team, the client organisation and among the project financiers, who are even less likely to have previous construction experience.

Other aspects of communication on construction projects could be supported by object oriented modelling. Many of these remain untested in a full project environment. An example would be the management of technical data between key members of the design team. If all were designing in an object oriented modelling environment, data would be attributed to an object which would then be available to the design team. A checking and clash detection exercise could then take place at the end of each working day to establish how data had been modified and what the results were.

Certainty

This last aspect of communication is important because it helps to eliminate uncertainty within the design process, in particular that introduced by incorrect data transfer throughout the design. This, in itself, is not the biggest advantage of the use of object oriented modelling. The major benefit that has accrued to those using the technology in advanced manufacturing is that it has automated their existing development and testing phase.

The terminology depicts a design process which has been carefully phased to allow testing of the concept in various forms from an initial sand or clay model through to an actual physical prototype. By modelling the concept in this way the user arrives at the point of manufacture with an almost 100% certainty that the concept works in practice and achieves the stated performance criteria to the satisfaction of the customer. With a large production run to follow, the expense of this design process is acceptable. For a one-off design the ability to model *in toto* is not possible. The prototype is the one that remains for posterity.

Using object oriented modelling throughout the design process would allow the concept of electronic prototyping to be used. The design process is remodelled to take into account the production at various gateways of an electronic model. As the design progresses so certainty about the particular components and the way they interact is increased. Just prior to construction, the uncertainty within the project should thus be limited to those uncontrollable external risks, such as the ground conditions or the weather. The product certainty may not be as close to 100% as in manufacturing but should certainly be in the region of 70–80%.

Other attributes that are deemed pertinent can also be built into this design process. For example, the appropriate software will allow the number of different components within the model to be reduced. In some instances the model will highlight small changes in grid pattern, for example, that will result in the number of different steel beams required being reduced. Likewise, buildability can also be checked to a greater degree.

In developing this approach to design using the concept of electronic proto-typing the result is a more tightly engineered product. It has an increased level of certainty not only in concept, by meeting the clients requirements, but also in the product itself, where costs have been engineered out by greater dimensional certainty and clash checking.

Coordination

There are two levels of coordination that can be improved by object oriented modelling. The first is at a scheduling level where, by linking the model to a construction programme, more accurate predictions of delivery dates and start on site dates can be achieved. In turn phase access and storage space can be assessed, as can the coordination of work teams in areas where space is small or there is a high hazard risk. Another advantage of using an object model is the ability to discuss the approach to a particular sequence of tasks with different work teams. Using the model allows a common understanding of the problem and therefore a more appropriate solution to be found.

The second aspect has been discussed earlier and is the coordination support afforded to the design team at all stages of the design process. One unusual example of this is the modelling of an existing building for refurbishment. If an existing structure is to remain, the intended new construction can be superimposed on the existing model and any problems of interference identified and rectified. Often in practice these sorts of problems can only be resolved on site, where they are costly to put right.

Major changes ahead – a quantum leap

There are a number of factors which will ensure that the UK construction industry cannot remain an island no matter how hard it may try to do so. The first is the free market within the EU. This factor may prove to be more subtle that it first appears. It is likely that multinational clients who are regular industry clients will develop advanced methods of design and manufacture/assembly with framework partners. This will be undertaken by them as a standard part of their drive for increased competitiveness. The capital cost and occupation costs of such clients will be subject to value for money exercises as part of company-wide continuous improvement programmes. The quest for competitive advantage will be driven by competition from within Europe, from the US and in particular the Asia–Pacific region.

One result of this will be that companies are likely to be increasingly globally aware. With current efforts to develop more rigorous benchmarking methodologies, international comparisons are likely to be both widely available and increasingly reliable. This will stimulate value for money initiatives and, as a result, will be a powerful engine for change. Those who take part in the construction procurement process, designers, component manufacturers and suppliers, planners, organisers and assemblers both need to understand and get to grips with supply chain management techniques and the concept of global purchasing now used so successfully in other industries.

Advanced IT, a proven means of delivering the change, is itself seeing dramatic breakthroughs in both hardware and in software. Since the early 1980s there has been a dramatic improvement in computer power, data storage and price. For example, Pentium technology PCs are considerably faster and more powerful than large mainframe computers of just a few years before. In addition, the new technology is more reliable.

Finally, it should be remembered that the PC became popular in the first place because it allowed users to work out their own ways of doing things, rather than having to choose from a menu of predetermined options decided by a remote committee or Stalinist IT department.

Although the most radical changes currently taking place in the software industry are with, for example, object oriented technology, there are a host of other developments, each of which will have a great effect on how computers are used. These include the increasingly powerful and all embracing Windows™-based operating systems and new interface management software that has the potential to link together any of a wide range of application packages in a unique user preferred way. It is clear that software is a booming industry. To an outsider it seems that it is controlled ever more tightly by corporate giants. In the past, IT innovation had been by one, two or three individuals, probably in their early 20s working in a parental garage. It appeared that the big software vendors had halted this.

The Internet is changing the rules to again favour the small inventors with the good idea. This new opening-up of the market is allowing dramatic progress in the development of software, particularly for the Internet. In 1995 Sun Microsystems, a workstation manufacturer, announced the missing link, a computer language designed explicitly for networks, which they named Java. As *The Economist* put it

It is a way of writing small programs, each designed to do a single task, that reside in the network until needed. When you click on a Java-enabled Web page, one or many of these 'applets' is downloaded to your machine, where it runs. It could be something as simple as an animation or as complicated as a mini-spreadsheet ... Whereas Web pages are static documents, Java applets are real programs that you can control like any other software. Once they are downloaded, they run locally on your PC, largely free of the network's delays. You request only the applets you need, which will be no bigger than they need to be to do their one job. When you are done with them they can be disposed of, or your PC can store them to save the downloading next time.

Best of all, you may not even notice they are there: content and software can be seamlessly merged. When you click on an analyst's company report, it seems only natural that the company's share price and latest financial results should pop up in a chart that can be manipulated, or that news headlines on a firm should scroll underneath. These are Java applets, working automatically in the background.

The Economist, 25 May 1996

All of this points the way to the future of computing for construction. Using Java technology on the Internet it will be possible to access and suck down dynamic information on, for example, manufacturer-supported routines for calculating steel sizes, project-specific window or glass size calculation routines, up to the minute stock lists and order times. If linked to object oriented technology, electronic catalogues containing not just dynamic information but also manufacturers' generic object models, say of a lift and the lift shaft it runs in, can be downloaded straight into your integrated project database and knowledge based model. For an industry that is notoriously fragmented, but reliant on accurate information and where coordination and management of information is a continuing issue, this Internet technology offers exciting possibilities in the short to medium term.

One 'quick win' idea for the construction industry, with its design team made up of numerous specialists, is the concept of the virtual design team. Used in other design-to-order industries, this concept has already been used in construction in the US with considerable success. How it works is well documented and references are given at the end of this publication. Such an approach has been shown to dramatically improve design coordination and productivity in the process industry, where in the US it is often known as *the 24 hour global design office* concept. At the root of this approach is the realisation that the geographical location of members of a design team is irrelevant. With a master project model and suitable design management tools, such as clash detection and specification/performance violation routines, designers in different time zones can work on their aspect of the design, post it electronically to the master model, where clash detection and other routines will be undertaken, before the new piece of design is accepted into the master project model.

When can object oriented modelling be used?

Since the early 1970s, systems thinking has fundamentally affected the way complicated things are designed, or complex organisations are viewed. It has revolutionised a range of disciplines from electronics, computer software, aerospace and air traffic control to the design of business organisations. The rewriting of physical geography by a team of Cambridge University researchers in the early 1970s facilitated considerable advances in the understanding of geographical systems.

Systems thinking allowed complex things to be mapped, to obtain a holistic perspective. The map could then be chopped up into a series of interconnected bits so that each bit could be examined as an entity. Such an approach allows practitioners to develop, in effect, road maps through a series of interrelated complex issues.

For many in the construction industry, the idea of mapping a flow of data around an organisation, or of mapping a project process, is still alien. To them the design process, or the architectural business process, is *The RIBA plan of work*, while the contractors' estimating process is a list of sequential activities (tasks) the estimator performs. For example, there have been attempts to look, from a systems perspective, at the flow of information around several contractors. In addition, several major clients have attempted to map parts of their business process, to aid a better understanding of the process and facilitate process improvements, in order to achieve industry best practice *en route* to world class. Although each of these projects have been judged successful by normal business/research criteria, the widespread use of such thinking and the clear benefits that accrue have not followed, largely for cultural reasons. In particular, the deeply held but erroneous belief that the construction industry *is different*, and that each architectural or civil engineering project is *so unique* that any commonality of system (if it exists) is so small as to be of little value, is a major barrier to improvement. However, as this publication shows, the differences between design-to-order and design-to-manufacture industries are starting to disappear for many types of projects where regular clients of the industry are prepared to standardise in a number of different ways to achieve capital and whole life cost reductions.

To date the industry has undertaken a large number of computer based initiatives, usually centred around a particular task or activity that is part, often a key part, of the traditional UK construction process – for example, scheduling, estimating, aspects of structural (steel or in situ concrete) design, aspects of HVAC design and working drawings. Almost all initiatives have been in isolated areas of expertise. With the use of process maps, the opportunity exists, firstly, to link all of these innovations together, to make each an incremental step in a larger change and, secondly, to allow fundamental process reengineering to take place to harness the benefits of the technology, rather than the automation of existing practice with only limited benefits of undertaking traditional processes.

Electronic prototypes – much more than visualisation

From childhood we learn to observe, live and work in three dimensions. Whether it be learning to catch a ball, ride a bike or play a field based team game. For a few, spatial ability remains low but for most it can be developed to very high levels with suitable technical training. Airline and helicopter pilots, architects, engineers and sportsmen can often achieve high levels of spatial ability, albeit in very different ways. However, for most non-technical people, drawings in two dimensions remain difficult to read and the projects that they illustrate are very difficult to imagine in three dimensions. For many clients and potential building users this is a severe handicap when negotiating with designers early in a project and too often it leads to nasty surprises and dissatisfaction at the hand-over/occupation stage.

Of potentially greater importance is that the client's skills, gained from many years of knowledge and expertise of working in a hospital, an airport terminal or a hotel, are not harnessed and, thus, only limited continuous improvement can take place. With the use of real 3D visualisation techniques, the greater realism allows the non-technical to relate their expectations, needs and experience/knowledge to design ideas and schemes. These benefits can accrue with electronic models or prototypes whether using advanced graphics such as Reflex or MicroStation, or simpler more crude images/illusions such as virtual reality.

However, such ideas demand a rethink of the traditional ways of working and the often latent but well honed traditional design or business processes, if the maximum benefit is to be gained from the new technology. One very clear feature of the success of a number of advanced manufacturers – such as Boeing (777 and 737–700 aircraft), Northrop (B-2 stealth bomber), Toyota (Lexus), Jaguar, PSA (Peugeot and Citroën) and Lotus cars – is that the technology has been used as part of a fundamental reworking of the design and production process, rather than using it as little more than a marketing tool. This criticism can be justly levelled at some users of the technology in construction, as indicated in reports prepared for the UK Government's Department of the Environment Construct IT initiative.

Benefits from object oriented modelling technology are possible in a number of key areas. Northrop in the late 1980s used knowledge based 3D CAD/CAM systems to design the very complex curves and angles required on the B-2 bomber and to achieve the very high level of manufacturing accuracy needed to maintain its stealth characteristics. The success of this new design technology

BOEING Phoenix Project

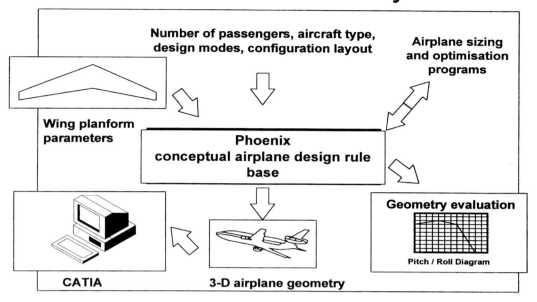

Number of passengers, aircraft type, design modes, configuration layout

Airplane sizing and optimisation programs

Wing planform parameters

Phoenix conceptual airplane design rule base

Geometry evaluation

Pitch / Roll Diagram

CATIA

3-D airplane geometry

Source: Boeing

meant that the B-2 became the first aircraft in history to go straight from the drawing board to manufacture without the need for physical prototypes. This process was then adopted and improved by Boeing for the 777 and the later 737–700.

Within Europe, a major research and development project has looked at the application of advanced IT to the engineering manufacturing process. Funded by the European Union under its Esprit programme, *Advanced information technology in design and manufacture* has involved partners from most of the leading industrial engineering groups. It identified five technical work packages and the topics each of them would investigate, namely

- product definition (layout system, concept model design, integration of electronic system design and development, and a simulation platform for product dynamics)

- product modelling (digital mock-up, enhanced CAD systems for component design fit and testing, interface CAD/CAE, product configuration management)

- manufacturing engineering (simulation platform of automation devices, integration of process planning activities, manufacturing on different production sites, management of manufacturing data)

- production control logistics support (flexible production and logistic processes, management of activities performed on shop floor level)

- information management (product data standards and STEP evaluation, information/data management, knowledge collection and utilisation, IT and telecommunication tools for concurrent engineering).

The project made considerable progress in the areas of business reengineering and the implementation of IT based advanced technology. Partners reported significant improvements in product design, component design, better component fit, faster assembly, better design and information coordination, and shorter times to market. In addition, they reported overall cost reduction through greater use of roll-over (standard) components and dramatically increased overall productivity (reduced man-hours per unit of car or manufactured product). There are now a number of clear benchmarked cases of overall improvement, sometimes dramatic, resulting from this approach in design-to-manufacture industry.

Design-to-order industries, with unique projects and issues such as those in construction, are also using this approach. For example, the petrochemical and process industries have harnessed this technology as a key part of total process reengineering. They are using the technology to model concepts at the business idea/opportunity stage. Later, it is used to develop a model of the facility built up from standard and other components, known as a *digital mock-up*. Although the emphasis and nature of much of the data will change between stages, when using object based technology the two models draw on common databases as well as specialist ones.

Several recent reports have demonstrated the need for this approach in construction. However, aspects of these reports, in particular the Latham report, have been misunderstood widely in the industry. Many, latching on to Latham's target of a 30% cost reduction in real terms and echoing important industry clients, have genuinely been unable to see how this could be achieved (using current practices) and have loudly denounced such comments! Yet there is clear evidence that such targets have been met and surpassed in other design-to-order industries. This has happened where the technologies described in this report have been embraced, where fundamental process reengineering has taken place, where major training programmes have helped key personnel develop new skills and attitudes, and where the senior managers concerned have adopted both a *can do* approach and a *continuous improvement* philosophy in their business.

Advanced information technology in design and manufacture

A major study funded by the European Union

This large ongoing study by a consortium of major European manufacturers began in 1992 with a mission to investigate how IT could be used to develop competitive advantage for European manufacturing.

Five work packages have been identified. Of particular interest to the construction industry is the work already completed in the development of digital mock-ups (DMUs) and the development of concurrent engineering process models and tools within the design and manufacturing industry of Europe. Work with DMUs has developed to a point where a methodology has been developed (company or project specific) on how to manage and visualise design data. This approach is now being widely and successfully used within the consortium.

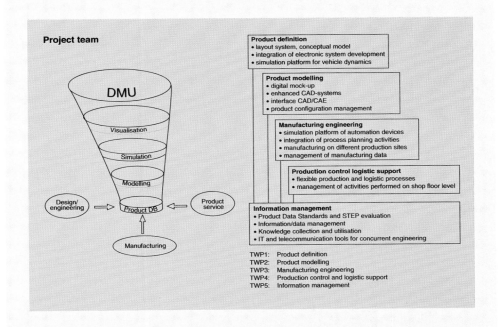

State of the art IT for concurrent engineering has been identified and methodologies for the design of concurrent engineering process models developed and used successfully in a range of companies and situations. So far, the work has overcome considerable scepticism in the manufacturing companies represented. Now that the concept has been demonstrated and deliverables identified, consortium members are showing remarkable levels of commitment to the project and to the private ones within individual companies that have resulted.

Consortium members include: AEG, Aerospatiale, Audi, BMW, British Aerospace, Daimler–Benz, Dassault, Fiat, Mercedes–Benz, PSA (Peugeot, Citroën), Renault, Rover, Lucas and Siemens.

Background

The Department of the Environment is sponsoring a research project at ACT-Reading under the title of *The transfer of advanced manufacturing (TAM) technology to construction* with industrial support from BAA plc and Concentra. The objective of this TAM project is to promote the transfer of advanced manufacturing technology into the construction industry. Central to this is the use of knowledge based engineering systems (KBE) to model prototypes prior to construction. These techniques are becoming increasingly popular in other manufacturing industries where they have proved to be both cost-effective and conducive to improved quality control.

The Schindler visualisation tool

Through discussions with directors and technical staff at lift manufacturer Schindler, it was established that one of the key weaknesses in their current sales to manufacturing process is the time required for the client to sign-off an order, committing themselves to a particular configuration. The basic requirements determining the size and performance of the lift could be quickly established, however it was common for late changes in the choice of finishes etc. to delay the final signing-off of the order. In some cases clients make changes that have major weight implications, e.g. from laminate to reconstituted stone finishes, and these would affect much of the mechanical specification.

The project consists of a pilot study to produce a proof of concept visualisation tool for the Schindler 300 series of lifts. With this tool the client sits with a salesman and generates their lift complete with car and doorway finishes, operating panel configurations etc. At the end of the process a detailed specification is displayed complete with fully rendered views of the car and doorway finishes which enable the client to see exactly how their chosen combination of finishes would look. They can then sign-off the order or take a hard copy away for the final approval of their directors.

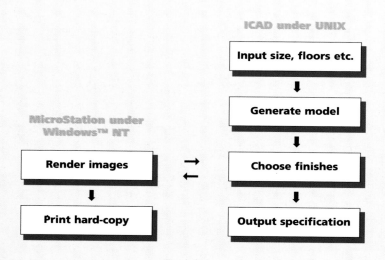

The 3D modelling approach

The system uses the ICAD KBE development environment to model the functionality of a lift car. Components are thus represented diagrammatically, but in a dimensionally correct manner, to facilitate output to a model visualisation device (MVD), MicroStation in this case. Output from the prototype is thus in two forms: a 3D block diagram giving the functional relationship between components to assist in communicating the engineering, and 3D rendered images produced through a separate MVD package as described above.

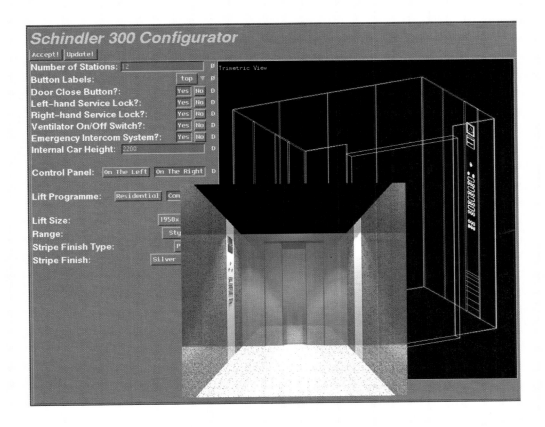

Future developments

Schindler are currently quite enthusiastic about the potential for developing this system and are currently evaluating the options for a much larger follow-on project, which would encompass the automated production of costs, builders work details and extension to other lift ranges.

Conclusions

The visualisation tool created for this study derives much of its value and potential for future development from the fact that it is far from merely a 3D virtual reality tool, but instead is a fully detailed engineering model which also has the capability to generate 3D visualisations. This means that any configuration options are automatically constrained by the engineering rules upon which the lift design is based, while in addition there is almost limitless potential for expanding the system to integrate with, for instance, drawing production and bill of materials generation.

Genesis project

Genesis is a BAA plc project based at the Heathrow Cargocentre, on the southern side of the airport. It involves the construction of a multi-storey car park in order to relocate staff parking to free up circulation areas. This project is being carried out in a unique manner. Dissatisfaction with traditional construction procurement has prompted BAA to manage projects under their own contractual procedures. Genesis has gone a step further by employing systems learnt from experience and research into the manufacturing industry, in order to adopt a potentially world class project management approach.

A decision was taken to use the Genesis project as the first trial for using the construction phasing detailed within BAA's new project process. BAA had already arrived at a possible plan for process improvement with construction projects, but it was felt that the best approach to implementing a new project process would be to allow the consultants appointed to agree mutually the details which were crucial and how they might be employed. During the first two and a half months of the Genesis project, the team members developed the processes and procedures. Project systems or tasks which operated within these processes were also identified. The focus was on establishing what factors were most important with respect to the specific problem of constructing a multi-storey car park at Heathrow Airport.

The most striking organisational difference seen on Genesis was the integrated core team or Project Executive Team (PET). The way the team operated as a single unit provided great benefits to the project and also the individual consultants. The project was further broken down into three 'delivery teams', each responsible for the detailed development of one of the identified project zones

- substructure and infrastructure
- car park and cores
- retail.

Each delivery team had two or three representatives from the PET and representatives from each of the appropriate supplier/contractor designers. Separate programmes were drawn up for each delivery team so that they essentially operated as separate units within the same project. Weekly meetings were held by each delivery team to examine progress and plan future work.

IT was identified from the outset as a key resource to be explored within Genesis. It was thought to have a large potential, but also that many learning issues, particularly in terms of organisation, needed to be overcome for full exploitation of its functionality. In addition, the practical requirements for implementation of IT needed to be examined, such as the provision of a personal computer for all project members.

The most effective information technology tools available to the project manager were

- integrated planning
- 3D modelling
- workshop culture
- central planning tool
- KBE systems
- cost management systems
- computer based management systems
- virtual reality
- bar-coding
- central project database.

From those evaluated, only the ones felt to offer a genuine benefit were selected. No tools were selected for their novelty value alone as the project was not intended as an IT test bed. The chosen tools were employed to varying degrees and at different periods of the project's duration. In particular, the use of interactive 3D modelling enabled the designers to have a far greater understanding of the design and how it would be assembled. Virtual reality visualisation of the building allowed an opportunity to visit the final product during its design, and to visualise the construction of the building. This was of benefit to the designers and to the client, who could see the end product early on.

Case studies

BDP
Cribbs Causeway Shopping Centre

Stanley Bragg

Alfred McAlpine

Kyle Stewart
Brindley Place

Fulcrum Integration Consultancy

Taylor Woodrow/BAA
Gatwick Airport North Terminal Domestic Facility

BAA/Laing
Heathrow Express tunnel

Tate and Hindle
Forbury Gardens

Cribbs Causeway Shopping Centre

In 1990 the Building Design Partnership (BDP) committed themselves to a clear IT strategy. They would dispense with the idea of a single inputter of design information at CAD terminals. Instead, everyone would have terminals for CAD and IT, using only a single (PC) platform. Within BDP there is a clear preference of the various professions for particular design packages. Interior designers, mechanical and electrical designers, and structural engineers use AutoCAD, while most of the architects use MicroStation.

BDP later became involved in the design of Cribbs Causeway, a large shopping centre development in a prime spot on the outskirts of Bristol. The client was keen to encourage the use of innovative but not prototypical technologies for the project. Two IT initiatives developed from this

- the use of advanced communications systems to manage the project
- the use of modelling.

BDP set up an office specifically for the project which had the appropriate ISDN communications links to enable the exchange of 2D schematics with others involved in the project including Gardiner and Theobold, Ove Arup and Bovis. To overcome the problems of intellectual property and uncontrolled manipulation of the 2D files, the concept of outfiles was used. These are files which are seen as representing a completed stage of working (i.e. the designer is not midway through changing it). The outfiles are then published or made available to the other companies involved in the project as soon as the particular design work has been incorporated.

Two approaches were taken with respect to modelling Cribbs Causeway. One complete walk-through model was produced but this was found to be too sketchy with respect to many of the details that were required by the client. The other approach involved the development of detailed photorealistic images using ModelView. This provided sufficient detail of natural lighting effects and the appearance of various materials (which were scanned into the model from samples). Also, to give an accurate impression of scale, people were incorporated into the finished images.

Why was the client concerned about the photorealism of the shopping centre images? The client recognised the effects that minor alterations in design can have on the appeal of the centre to potential shop tenants. The prospective tenants, in turn, recognise the subtle effects that building design can have on shopper behaviour. A shop unit with low visibility and low levels of passing pedestrian traffic is likely to achieve very low sales. The essence of shop unit selection is position and visibility. The Cribbs Causeway development is typical of a blueprint applied in almost all shopping centres. This involves the use of anchor retailers who are large and established chains such as Marks & Spencer and John Lewis. The smaller retailers rely on the presence of these larger outlets to attract custom. By the careful distribution of these units between the two anchor stores passing traffic is encouraged to visit these smaller outlets. Between the two anchor retailers there is also a minor anchor store, which would normally have a more specialised range of products, such as a large pharmaceutical outlet.

Another important reason for modelling the project in detail is the need to convince retail outlets of the cocktail of outlets which is being considered. A poor mix, with undesirable or competing outlets, is likely to deter potential tenants. The model is, therefore, used to show tenants the exact appearance of the shop front seen from various angles and levels. Potential tenants are also interested to know the visibility of the anchor stores at each point. A concealed anchor store is likely to deter shoppers from passing a shop unit.

Conclusions

Improved design can not only reduce costs, it can also increase revenue. Photorealistic modelling of Cribbs Causeway enabled the designer to capture feedback from the client about many small design changes which would enhance the desirability of the facility. Furthermore, it enabled the client to convince potential tenants of the ambience and cultural mix that would exist in a facility which was yet to be constructed. Without the use of 3D modelling it is difficult to envisage how this might have been achieved. Compared with the first layout of the scheme the effect of design changes as a result of using the 3D model was to increase the value of the scheme by 50%. This statistic provides a very convincing argument for the use of 3D modelling and modelling in general.

Acknowledgement

Nick Terry, BDP, Gresse Street, London

Stanley Bragg is an architectural practice, established 50 years ago, in Colchester. The practice was involved with object modelling at an early stage of its evolution, with the introduction of the RUCAPS CAD system in 1982. This involved a considerable investment and there was a crisis point when the pursuit of an object modelling approach seemed to be like one financial black hole. To recoup their object modelling investment and to set it straight on the rails Andrew Ling, the systems manager, introduced more rigorous systems surrounding the use of the object modelling system. This involved the use of formal structures, methods, training and familiarisation – essentially, the management surrounding the use of such an innovative approach. Ling acknowledges his good fortune in receiving the backing of the firm's directors who were willing to adopt these changes and to continue their commitment to object modelling.

Company culture

Today, they are routinely using object modelling as the core for their entire business. They have developed a library of components for use with the Sonata object modelling system which they have been continually developing for the last seven years. Ling is emphatic about not using object modelling as a 'bureau' which serves the practice. All design is undertaken by architects using the object modelling environment directly – not data entry operators. All the workstations are operating off a local area network enabling numerous designers to work on the same model concurrently (multi-project access or MPA). Conventional CAD such as AutoCAD is only used for the purposes of file transfer from incoming consultants and contractors. Much emphasis is placed on training and on integrating object modelling into the company's culture. Regular workshops take place and newcomers, be they 50 year olds or 20 year olds, are trained from the word 'go'.

Stanley Bragg often use Sonata to create a full 3D model of a scheme prior to work stage D, at which time they present a fully rendered model and various proposals. Ling acknowledges that this demands more effort up front and there is inevitably a degree of risk if the firm fails to win the work. But if the project proceeds, the return far outweighs the extra effort as working drawings are generated directly from the 3D model.

File management

From a system manager's point of view Ling says that drawing management is easy. Producing drawings from a central model always provides the most up to date information for distribution. In conventional file based systems, it is notoriously difficult to manage drawings as part of a continuous project process. With Sonata, using parametrics, it is possible to model different levels of information at different stages. Thus, the same line for a wall can be shown as a simple schematic for a planner, or as a detailed cross-section for detailed design. Another advantage, in contrast to conventional CAD, is the ability to present drawings at any desired scale.

Layering

Layering is also easily dealt with using object modelling. The object modelling system's intelligence means that the designer cannot inadvertently enter information on the wrong layers and it is possible to impose a strict layering convention. For example, wall information can only be entered on wall layers 230–240. Objects placed on the wrong layer (e.g. ducting on the drainage layer) are rejected. In contrast, CAD systems will happily accept walls on drainage layers and many other layers: Ling comments that often he receives CAD models with trees appearing on half a dozen layers!

From Ling's experience 'layer management ceases to be an issue'.

Stanley Bragg are keen to see a move towards the idea of a common object model which is shared by all the participating organisations on a project. They have experienced difficulty in finding the right people to work with, in terms of structural engineers and services engineers, who have expertise with 3D systems. However, they are now working with a number of specialists who are able to share a common approach to object modelling, and among others they are working with Fulcrum, an M&E specialist. Recently they have experimented with the use of ISDN lines to make the object model integrate with these M&E specialists in Reigate, Surrey.

Ling continues to evangelise about the future of object modelling in his company. He remarks with a wry smile about developers who pay for fit-outs of entire floors in order to attract tenants. He believes that we (the construction industry) are going through 'quite a transition and some people are in for a shock!'

Acknowledgement

Andrew Ling, Stanley Bragg, Abbeygate One, 8 Whitewell, Colchester, CO2 7DF

McAlpine have been heavily involved in the development of a cost estimation system (in conjunction with Salford University and Engineering Technology Ltd). Using Reflex, a 3D model is generated to establish structural, mechanical and electrical costs developed from information on usable floor area. This addresses a particular need in the industry, to provide reliable cost information at an early stage, fulfilling the needs of the project quantity surveyor (PQS). It develops a model from only outline ideas by applying a collection of consistent assumptions within the expert system. Estimating accuracy using this system is approximately 5–10% (comparable to the accuracy achieved by conventional estimating methods). This, in many ways, is perhaps not surprising, since it is the information upon which the estimation is based that determines the accuracy more than the estimating procedure itself. In a sense, the expert system is only automating what a PQS would otherwise do. The interesting challenge was to improve the accuracy of the system by providing more detail at the concept stage. However, this proved to be problematic since it is difficult to carry out calculations on elements each of which is at different points of detailing. Validation of the model's accuracy was impeded by the lack of reliable cost data on completed buildings. For situations where the predicted cost was markedly at odds with the actual cost (up to 17% in one case) the principle reason was the omission of major items such as air-conditioning, when the model had suggested that air-conditioning was required. This leads on to another point concerning the model's lack of awareness about *flip points*. The cost estimator will usually be sensitised to areas of the design where minor changes can result in major savings. For example, where minor changes to orientation might eliminate the need for air-conditioning. However, the computer model does not possess this awareness about 'flip points'. Another interesting finding was how many variables such as ceiling height could be altered without substantially affecting total cost, since M&E costs (dry risers and sprinklers) increased as structural costs decreased.

Another modelling approach, which is being pursued by McAlpine, involves the identification of key construction modules. This approach takes the manufacturing model as a guide. In manufacturing, designs are rarely bespoke but reuse a number of solutions which have been tried and tested before. This approach is entirely at odds with the construction industry. The advantage of using a proven module is that

- you know it works

- you know how much it costs

- you know how long it takes to build

- you know how it interfaces with other systems.

Also, using modules, it is possible to generate a detailed model using only limited information. An example which has been exhaustively explored by McAlpine is the design of steel framed industrial buildings. In principle, it is possible to design an industrial building of any size or height. But in practice a span of 18 m is a common standard width, although values such as 20 m and 17 m might also be chosen. Such varied values might be chosen because there are envisaged economies. However, the knock-on effect of additional design effort and structural analysis will more than off-set this need for flexibility of portal frame sizing. A major benefit of using standard dimensions is the opportunity for improved supply from manufacturers. Inventories can be improved and availability increased, reducing construction lead times.

This concept of modules has been applied to all other elements of an industrial building, including the type of entrance configuration at the gable ends. Variations of segmented, roller shutter and conventional doors are capable of being plugged into the design in a manner which is known not to cause conflicts in terms of dimensions or function.

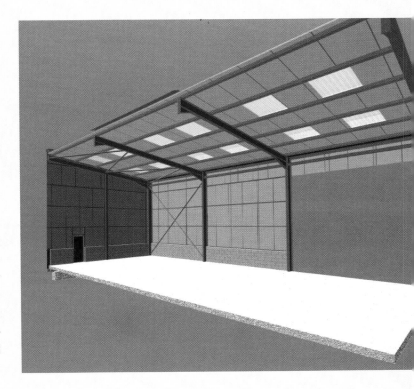

Using the modularisation approach McAlpine have been able to use Reflex to produce a fully dimensioned and detailed factory building based simply on the key design parameters. Knowing the portal frame width and height, the program is able to determine all aspects of the structural elements, cladding and roofing.

McAlpine are keen to stress that this approach is not system building often associated with inferior work: it is a modular design. Clearly, the logic of reusing proven designs is compelling. The push towards object modelling is likely to demand a greater use of standard design solutions. Inevitably, for some people in the design sector who make a living from producing designs from scratch, this approach may represent a threat. The challenge for object modelling developers is to produce design modules which are sufficiently flexible to offer an *individual* design solution for all building types.

Acknowledgement

John Hampson, Alfred McAlpine House, Nutsford, Nr Manchester

Kyle Stewart have been established in the design and construction market since 1953 and now employ a multi-disciplinary design team numbering over 130 employees. They recognise the importance of a common modelling approach and have been actively involved in redefining IT strategies for the industry. The project based approach to design and their commitment to partnering makes IT integration a key issue. The demands for technical communication and a single point of focus have resulted in the development of a sophisticated design system involving the integration of CAD, project and visualisation data.

The key programs which are used in the design process are

• CYMAP: a program for modelling flow systems in buildings

• A 3D ductwork design program which is able to obtain design data from CYMAP

• MICRODRAIN: a program used to design drainage systems compliant with health and safety legislation

• INTEGER: a structural design program

• PowerProject: a generic scheduling program

• Sonata: an object based CAD system (predecessor of Reflex) capable of interpreting CAD information transferred in the standard exchange form DXF. Sonata can also use time-evented information obtained from data generated in PowerProject.

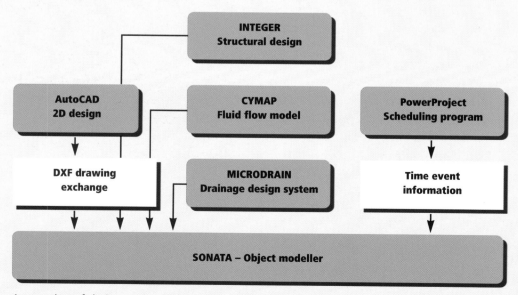

Integration of design tools with object modelling

Brindley Place

Brindley Place, located in the heart of Birmingham, spans 17 acres and comprises a mix of office buildings, private accommodation and a complex of bars, restaurants and shops.

Argent are the key developer involved in the project and British Telecom signed up as a tenant for No. 5 as a flagship regional headquarters. Argent, as a developer, were keen to see the use of IT developed as a major communications aid driven by cost and quality considerations, and wanted to see IT used as a device for integrating the project team, providing links with the various disciplines and facilitating the use of value engineering.

An undoubted benefit of object modelling in the eyes of Kyle Stewart is the ability to 'deliver precisely what you say you are going to deliver' and IT enabled the project manager of the Brindley Place project to be more confident about the way the job would run. The availability of information up front has enabled co-ordination issues to be addressed before arriving on site enabling the work to run smoothly.

Argent were able to accommodate the demands of the tenants in the pre-let office space at an earlier stage.

Acknowledgement

Colin Delaney, Kyle Stewart, Merit house, Edgware Road, Colindale, London

Fulcrum Integration Consultancy

Fulcrum Integration Consultancy is one of a new breed of specialist consultant, who have been able to undergo rapid growth as small start companies founded on the use of knowledge based technology to enable design, development and integration. Fulcrum Integration Consultancy only began trading as recently as October 1995 and have already secured contracts to work on the Royal Opera House and the parliamentary buildings in the UK. Over the first six months much of their time was spent writing their own model library specifically for plant layout and mechanical and electrical services. While making use of various CAD systems including AutoCAD, MicroStation and 3D Studio, the central hub of their design is based on Sonata.

Multi-project access is achieved using Sonata so that up to five or six people are able to work on the same layers of a model at the same time. Technical difficulties have been encountered in accommodating more than this, so that on particular rush jobs it has been necessary to work in shifts to make use of the system.

By allocating the various M&E design specialists to input specific elements of the model Fulcrum Integration Consultancy has been able to minimise the learning curve involved in their work. Work is segregated into toilet core, riser shaft, plant room and general floor arrangement design. By the use of separate layers with the provision of match lines it is possible to provide continuity between the various design solutions in the model so that there are no inconsistencies at the interface of each of the areas of work. Obscuration, whereby parts of detailing can be hidden, provides a method for making sense of a complex system.

Clash detection is an important consideration in any M&E solution. With the use of object modelling, Fulcrum Integration Consultancy is able to identify clash problems. This is carried out as a staged process; when only 50% of the design has been carried out, the clash detection programme is run to identify elements of the design where more than one element occupies the same point in space; notes are then taken and further clash detection is carried out on completion of the design detail. With the ever increasing importance of tolerances, clash detection is becoming a fundamentally important tool. For the work they are carrying out for the new parliamentary building, the under-floor services (ductwork, piping, air conditioning, cable trays) fit in very tight building tolerances of 2–5 mm as part of a modular system. If fire protection is required, the dimensions of components are changed, if only by a few milli-metres. Fulcrum Integration Consultancy has gone to some effort to produce a model library capable of accommodating fire protection characteristics.

For the future, Fulcrum Integration Consultancy is committed to multi-service projects, with various specialist firms (architects, civil engineers, M&E consultants) buying into the same central model. In this way they believe it is possible to fix the design earlier and ensure that the building is coordinated. Their ability to manage the flow of information within project programme deadlines complements and enhances the design development process.

Acknowledgement

Chris Penn, Fulcrum, Albion House, Albion Road, Reigate, Surrey RH2 7JY

Taylor Woodrow / BAA

Taylor Woodrow decided three years ago to develop their own object modelling potential based on the Sonata, subsequently Reflex, system. They recognised that this would entail a substantial investment in both IT hardware and people, which to date is in excess of £1 million.

A director of Taylor Woodrow was given responsibility for overseeing the development of the system and a manager appointed from the construction staff. A knowledge base of parametrics for building materials and processes was established in a confidential in-company library.

The training strategy was adopted and bright young construction graduates and professionals were recruited who knew something about construction and were highly computer literate. A new department was created comprising eight to ten operators, a project leader who is a very computer literate architect, and a departmental manager, all reporting to the designated director.

It took the operators around two months to understand the system sufficiently to be able to make a useful contribution to modelling project information. It was two years before Taylor Woodrow were able to develop the object modelling system to a level where it became a tool which they could use with clients and thus obtain competitive advantage.

The enthusiasm of marketing staff in articulating and promoting the usefulness of the system to clients is seen as a vital ingredient in the successful commercial exploitation of object modelling by Taylor Woodrow. Examples of its use on projects such as the Royal Albert Hall refurbishment project have been cited elsewhere.

Gatwick Airport North Terminal Domestic Facility

Gatwick Airport, which is owned and managed by BAA, is the second largest airport in the UK, with both a north and south terminal. The requirement for a new domestic facility at Gatwick Airport arose from the take-over of Dan Air by British Airways in 1992. Up until that point Dan Air had operated in the South Terminal and British Airways had operated exclusively in the North Terminal. With the need for rationalisation, it became evident that the two separate facilities would need to combine and be sited at a single terminal. The decision was taken to relocate the domestic facility previously used by Dan Air to the North Terminal. This would greatly reduce turn-round time on British Airways flights and enable aircraft to switch from domestic to international flights and vice versa, thus achieving greater aircraft utilisation.

In the intervening time it was necessary to construct a temporary domestic facility quickly, using prefabricated units at the North Terminal.

Design problem

The logistics of combining a domestic and international facility at a terminal are particularly demanding. Domestic passengers require a separate entrance to international passengers. Domestic passengers have only to undergo security checks, while departing international passengers must also pass through security and immigration, and arriving international passengers must, in addition, pass through customs. Careful segregation of these flows of traffic is required to ensure that control measures are enforced effectively. Another consideration is the distance that passengers are expected to walk going from the terminal to the boarding

gate by way of the walkways and piers. Domestic passengers expect a much faster boarding period with minimal walking distance, particularly as many domestic flights are having to compete with alternative modes of internal transport such as rail.

Part of the solution was to construct a circulation building which would maintain the segregation between the three flows of traffic. However, given the design of the existing facility, the circulation area itself would need to overcome complex geometrical problems so that the domestic traffic was able to drop down to the same level as one international route and below another. The circulation building would effectively allow the existing routes to be pulled apart and thread the domestic route in between them.

This project was seen as an ideal candidate for the use of 3D modelling. While there was no problem on the part of the project management team at BAA in understanding what was involved in the proposed solution, there were many other people who had to be persuaded of the project's merits. In the immediate term, it was the BAA Board, faced with many competing proposals, who needed to be convinced.

Process model

The North Terminal Domestic Facility project was innovative in two respects. Not only was 3D modelling used, but it was integral to the process adopted by BAA. This integrated process is called the *project process model*. It identifies key stages and processes in the evolution of a project from inception to operation and maintenance. Clear definition of these stages allows a consistent approach and ensures that projects are properly controlled. There are various milestones common to all projects which are termed *gateways* in the project process model. These gateways provide specific points in time for formal evaluation and approval. One of these gateways, following work stage D, concerns the consolidation of major design elements. It was at this particular

stage that 3D modelling was used to undertake a formal evaluation and elicit formal approval from the Capital Projects Committee at BAA.

The evaluation involved engineering and maintenance staff as well as many stakeholders who had not normally been involved at this stage. Users of the facility including airlines, handling agents, control authorities (including Her Majesty's Customs and Excise and the Department of Transport Police) were all able to contribute to the design process. Much of this occurred in the few days running up to D-Day proper when the final approval would take place. During this run-in numerous workshops were undertaken involving the various stakeholders. Plans and drawings were used extensively in these workshops, although 3D modelling was not used until D-Day. At this final evaluation and approval milestone, all the representatives of the various stakeholders were assembled in a single meeting room. The meeting was separated into two discrete sessions. The first session culminated in the signing of a document by all participants confirming that they understood the project and would commit themselves to completing the project without any changes in the design. The second session, in the afternoon, ended with the signing of a similar document which confirmed that all the stakeholders understood their own roles and responsibilities in the project and the customer experience arising from it. The earlier presentation thus focused on the design of the facility (led by Taywood) while the latter focused on the ground movements and scheduling (led by Schal).

For the first session a 3D model of the circulation building was developed and shown to the audience using a large display screen. The model itself was produced using Reflex, an object modelling package, making use of architectural drawings as input, to produce a rendered version of a wire-frame model. The model was presented as a fly-through and as a series of key still images, focusing particularly on the view of the passenger. There was no attempt to adjust the model during the course of the discussion; essentially, it was assumed that the stakeholders would have articulated their ideas during the workshops. In the workshops, Customs authorities identified inadequacies in the layout which would present a potential security breach owing to the overlap of domestic and international passengers at a certain point. Some of these design faults, gone unnoticed, would undoubtedly have resulted in costly rework, if such a mechanism had not been used to elicit these views.

Public relations

Another use of the 3D model was in relation to the temporary domestic facility which was in place for several months while the work on the new circulation building was undertaken. The fly-through generated from the 3D model was displayed in the temporary facility to passengers awaiting flights. Informing the customers about the temporary nature of the existing facility and the improvements which could be expected was an important undertaking since many of the passengers on the domestic flights are likely to be repeat customers. Ground staff at the other receiving internal airports (Manchester and Aberdeen) also made use of the model to explain to passengers what they could expect in the near future. It was also used as part of the package contractor induction.

The BAA experience

The success of object modelling by BAA can be attributed to a number of key policies, namely

- use of a consistent project process enabling proper project control

- use of standardised components

- use of framework agreements to develop close working relationships with a small number of suppliers

- use of concurrent engineering involving suppliers as part of integrated teams

- commitment to preplanning of fabrication and construction.

These policies justify and enable object modelling and 3D visualisation. What is also evident in its successful implementation is the commitment of the client.

Acknowledgements

Clive Coleman, BAA, Heathrow
Martin Fox, BAA, Gatwick

Laing have taken a restrained approach to investment in object modelling, and are committed to industry wide solutions which are being developed by the International Association of Interoperability. Furthermore, Laing are keen to ensure that any object modelling work which is undertaken can be carried out using low-cost computer hardware (Windows NT). They believe that this will make object modelling accessible to a larger part of the organisation. Object modelling is being undertaken in a rudimentary way using AutoCAD currently, and 3D visualisation is achieved using 3D Studio for off-line processing and SuperScape for real-time interaction.

Heathrow Express tunnel

The 3D visualisation of the Heathrow Express tunnel shows many of the benefits of 3D object modelling. It also shows what can be achieved using low-cost computing power to produce a convincing virtual environment.

Railtrack commissioned Laing to provide a 3D modelling system to help in identifying the correct positions to locate rail signals along the length of the Heathrow Express tunnel, which is 15 km in length. The 3D modelling effort undertaken by Laing addresses a specific problem: that of signal siting. Siting of train signals has traditionally been as much an art as a science, carried out by trained and experienced siting engineers. The driver of the train must be able to see a signal in view at all times. However, it is difficult to ascertain at the planning stage whether signals will be visible to the driver or not. When overground signals are being sited, the conventional approach is to undertake a manual survey on the ground. Siting engineers experiment with positioning dummy signals at various points along the length of intended track. The dummy signal will then be approached at appropriate speeds to establish whether it remains visible. If the signal is not visible it is repositioned and further tests are undertaken.

The 3D modelling approach

The manual surveying solution to the signal siting problem is time consuming and costly in terms of man-hours. In the case of underground rail tracks, the exercise itself can only be undertaken once the tunnel has been excavated and any objects which are likely to obstruct the line of sight are in place. The appeal of the 3D modelling approach is that the positioning of signals can be decided during the main design stage, prior to tunnel completion. Not only does the computer visualisation approach reduce man-hour costs, but it also reduces the total project time. This is because the siting survey does not have to be undertaken at the very end of the project, which would lead to a significant increase in total project time and subsequent loss of revenue.

There is a considerable amount of modelling effort involved in producing a driver's-eye view of the tunnel. Some of the parameters which have to be considered include

- tunnel coordination information (vertical and horizontal)

- tunnel cross-sections

- track position

- track camber requirements

- position of electric cable tensioners

- speed of train.

Curvature of the track, both horizontally and vertically, will affect the visibility of signals as they become obscured by bends in the tunnel. The camber of the track can also be important, since the driver's view becomes slanted as the

train tilts, often concealing the signal itself. The dynamic envelope is the cross-sectional area occupied by the train along its length, which also includes the area due to vibration. This increases as the velocity of the train increases. In turn, this may affect the options for positioning the train signals as do obstacles such as cable tensioners, piping and walkways.

This information is provided by various outside specialists (civil engineers, architects, track designers etc.) as conventional alphanumeric (ASCII) files. All this information is then combined into a single database linked to AutoCAD. AutoCAD is then used to produce wire-frame cross-section views or surface models. Finally, the AutoCAD views are converted into fully rendered and textured images using 3D Studio. These images need to account for the effects of dust within the tunnel which can reduce signal visibility.

Interpreting the 3D model

The presence of a 3D model does not displace the role of the siting engineer who still needs to make judgements about the suitability of the signal locations based on the simulation. If the signal positions are deemed unsuitable, the signals are then repositioned in the model and the 3D visualisation is repeated for that particular length of tunnel. The siting engineer is concerned with two parameters

- the approach view

- the uninterrupted view.

The approach view is the critical view from which point the signal needs to remain visible, allowing for momentary obstructions by objects in the tunnel. The uninterrupted view is the critical view from which point the signal needs to remain entirely visible without obstruction. With these two views the siting engineer is able to make an informed decision about where to locate the signals and will repeat this process for the many hundreds of signals which will be used through the length of the tunnel.

Conclusions

This highly focused application of 3D modelling shows how it can be used to help in critical design decisions. The particular benefits of using 3D modelling for this particular problem are

- the substitution of a labour intensive on-site surveying exercise with a faster visualisation method which provides the expert siting engineer with more immediate feedback

- the reduction in total project time for tunnel completion by eliminating what was a critical task on the project programme.

The case study also shows visualisation issues that are applicable to conventional building design. Static images often fail to show the impact a design can have on an observer. Movement, viewing point, variations in lighting, presence of obstacles, can all have an unanticipated impact. Only by mimicking this dynamic experience can aesthetic, functional and safety issues which are so critical to the client, be addressed. In the case of the siting problem, the effect of ignoring these dynamic interactions could lead to flawed design decisions with possibly fatal consequences.

Acknowledgement

Mervyn Richards, Laing, Page Street, Mill Hill, London

Tate and Hindle

Tate and Hindle are an architectural practice located in the West End of London with a team of 17 designers. Partner Andrew Tate believes they are now 'on the right side of a polarisation' which has left some architectural firms well ahead of many others who are left in an IT wilderness. Like many practices of their size, they have had to choose an IT solution which is affordable and proven. AutoCAD LT is the main drawing tool, used in conjunction with 3D Studio as the rendering and visualisation tool. Using a system capable of producing visualisation tools and 2D CAD, they feel that they can achieve a greater level of quality assurance than would otherwise be the case. They are aware of the limitations of their current system, since it does not allow them to move from concept modelling to production drawings in a continuous manner. Moreover, the issue of human error needs to be carefully addressed. They believe that object modelling systems could be a long-term solution to these problems, although the costs of hardware, software and development are prohibitive for typical design firms of their size. However, the move towards systems running on affordable PC platforms may provide the answer to this.

Peter Martin, Tate and Hindle's systems manager, believes that many architects welcome moves towards IT innovation, clients are perhaps only occasionally interested and 'contractors tend to go with the flow'. One of the stumbling blocks he perceives in relation to a common 3D model is that the construction industry involves many segregated disciplines, with each profession carrying their own professional indemnity and insurance considerations. This results in a communication barrier which is unlikely to disappear with the introduction of a common object model. Another difficulty he sees is that the parameters which a quantity surveyor is interested in often have little commonality with those which the architect is interested in, so the models themselves would tend to evolve as two processes.

Tate and Hindle tend to recruit people that have reasonable computer experience, but they were willing to train people if necessary. The important thing as designers is that they understand how to communicate design information.

Forbury Gardens

An example of how 3D visualisation has helped Tate and Hindle to communicate with planners and clients is an office development for a financial services company adjacent to a historic garden, Forbury Gardens, in the centre of Reading. One of the obstacles to the development was the planning approval. To help with communicating the scheme a photomontage was produced using 3D Studio images. Its presentation aroused so much interest in the planning department of the local authority that the presentation room was filled by people from the planning department curious to see the fly-through.

The use of 3D visualisation saved a lot of talking, since the clarity of the scheme obviated the need for this.

One issue which the planners were concerned about was the public right of way which was to be accommodated by a high narrow archway through the building. This frequently used thoroughfare is important to many pedestrians and the planners expressed concern about the dark and dangerous (if only perceived) appearance of the walkway. By redesigning the archway, Tate and Hindle were then able to reproduce the altered walkway in a 3D visualisation and convince the planners that it worked.

Short-term developments

One of the areas in which Tate and Hindle are heavily involved is concept design. In order to use visualisation tools they need sufficient information to make the images convincing. However, in order to do this, they may need to undertake excessive amounts of detailed design work. They are keen to adopt systems that will allow them to produce convincing 3D images, based on manufacturers' information, to help with giving a sense of workplace design which is important even at concept stage. Some lighting manufacturers now provide electronic information that enables 3D Studio to reproduce the lighting characteristics of a particular fitting. Another approach has been to produce the desired lighting effect using 3D Studio and request from a manufacturer suggestions for light fittings which might reproduce such an effect. Furniture manufacturers such as Steelcase produce 3D objects which can be readily assimilated into a 3D Studio image, which again can liberate the effort of designers at the concept stage. Martin believes the release of 3D Studio Max, which is object oriented (but which is limited to its own internal system), could help in managing the evolution of 3D images.

Acknowledgement

Andrew Tate, Tate and Hindle, Ramalese Buildings, 215 Oxford Street, London

Chapter 6

How will object oriented modelling become part of the industry?

What are the implications of object oriented modelling?

Earlier chapters showed how object modelling could change the way in which construction projects are carried out, giving the industry dual benefits of engineering excellence and dramatically improved communication with its customers. The previous chapter gave examples of a number of case studies where object modelling is being used in a variety of ways on actual projects and within companies.

Real benefits are accruing from these uses but their effects are relatively local. If object oriented modelling is to help shape the lean, efficient, creative, dynamic construction industry of the new millennium then its use needs to become widespread. This chapter investigates how this might come about through identifying two key drivers for change: the client and the individual firm. The longer-term implications for industry are examined and some consideration given to other implications, such as the role of training and research in the development of the application of object modelling within the industry.

The client as a driver for change

The first of the two key drivers is the client. Why would clients be interested in the development of object modelling? It could be argued that simply having the project concept conveyed to them by their design teams as 3D models would suffice. It is true that for many clients in the construction industry there is little need for they themselves to become directly involved with object oriented modelling. There is, however, one type of client for whom it could be particularly useful: frequent users of the construction industry who own and operate their own facilities. It is useful to them for two reasons.

First, because each owns a number of similar types of buildings, e.g. office blocks or airport terminals, they have a range of knowledge about how each building operates and what physical form allows it to operate in an optimum way. This knowledge is very important in the construction process because it can speed up the inception and concept phases of a project. The project process during these early stages involves a high degree of iteration between client and designer as the client determines the alternative business cases and the designer seeks to translate these into building solutions. A client who can furnish the design team with information gained from years of operating experience about the relationship between the business operation and the spatial requirements of the physical building can have two effects. The time taken to arrive at appropriate solutions can be dramatically reduced while the likelihood that the preferred solution will support the business in an optimum way is increased.

It is, however, the collection and analysis of the operators' experience that is difficult. The ability to store this knowledge and reuse it from project to project is generally low. Object modelling allows the long-term experience of operating the particular building type to be reused on new projects.

Secondly, it is important for this type of client because it allows them to remain in control of the project process. Clients of this nature usually have in-house project organisations to manage the workload of new projects. But the number and frequency of the projects make it difficult for consistency to be achieved across projects, especially where different external consultants are being used who each carry out project work in their own way. Consistent value for money is, therefore, difficult to ensure. By using a common tool, such as object oriented modelling, consistency is achieved both in terms of the quality of the building and in financial certainty.

The clinching thought is that knowledge is power and to a great degree the client who owns the model is in control since all information about the project, whether to do with physical aspects or financial, will reside in one place. Project team members will interact with the model, adding their skills to the project through the medium of the model. Since there will be similar repeat projects for the client the early investment and learning experiences will reap increasing benefits over a period a time. This would not be so for a one-off client.

There is also an advantage in streamlining the in-house project management overhead since object modelling will provide the tool for more effective management of both the process and the product itself for all the reasons discussed earlier. By using object modelling the client gets a greater ability to test out alternative processes and subsequent building solutions during the concept stage, thus arriving at a better solution to the business need.

Further possibilities also occur for the costing of new design, since a wealth of historical cost will be available for pricing of alternative designs. These can be attached to the objects as cost attributes. In this way it could be possible to deliver alternative design solutions with whole life costs attached.

It can be seen, therefore, that for an owner operator with an ongoing building programme the introduction of object modelling can influence the success of both the business and the construction projects themselves. But in reality, a great many construction clients are first time venturers into the construction market place. How could they influence the industry to use object modelling?

Clients who are experiencing the construction industry for the first time complain of a lack of control. The process is taken out of their hands and they feel vulnerable because of their lack of knowledge and understanding. This is exacerbated when construction professionals ask for decisions and information that the client cannot readily give. The key to making the construction industry a low risk solution to a business problem is communication. As identified earlier, communication is greatly supported by object modelling. Customers can more easily relate to 3D models because they are already familiar with them in their everyday world. They can, therefore, trust their own judgement about, for example, the combination of colour, light and space, because they are represented in a familiar way. In this manner, the customer's confidence in the building design can be built up.

These comments are based upon the applications of object modelling that are known about now. In time, as greater familiarity is achieved with object modelling in the construction workplace, other applications will no doubt be developed which may increase the benefit to new clients.

The individual firm as a driver for change

The owner/operator will be a major driver for the increased use of object modelling in the construction industry. The second, equally important, driver is the individual firms that make up the vast numbers of project teams. In the spirit of continuous improvement, these firms must consider new tech nologies as a key to achieving and maintaining their competitive advantage, and as a marketing tool to create new business. Until recently, the risks associated with object oriented modelling, both financial and technical, were too great for it to be considered as a potential area for investment. Recent developments have resulted in the supporting technology becoming cheaper and more readily available. As illustrated in chapter 5 there is a very real potential for a range of construction firms, from designers to specialist trade contractors, to consider how the services they deliver to clients could be dramatically improved by the introduction of object modelling.

The driving force behind the uptake of object modelling at the company level is its ability to improve the elements described earlier such as communication, creativity and certainty. To do this requires an in-depth understanding of where the competitive advantage lies for the individual company. Is it in the ability to help first time clients come to understand and to know what it is they need out of a building to support their business? If this is the case then the creativity and visualisation aspects of object modelling become more dominant. Is it the ability to react quickly to a retail client who wants to fast track the construction of a new store to meet a critical opening date at Christmas? In which case the ability to have a standardised approach tailored quickly to fit the specific project is important. This can be achieved by focusing the ability of object modelling to minimise the pre-construction phase. For site based companies perhaps, it is the ability to use visualisation to motivate and build an efficient team to achieve faster construction that is vital. The key issue for maximising the potential of object modelling is to identify the critical business factors that it will support.

The identification of these critical business factors and the subsequent application of object modelling to facilitate them allows a firm to develop a market niche. For example, an architectural practice with specialist airport or harbour knowledge could capture its expertise and develop a process which it can market to clients requiring work in the appropriate field. The concept of embodying the knowledge is proven. The individual firm must consider the context in which it wishes to use it and develop it to suit.

The approach for firms of this nature needs to be one of carefully analysing how and where they wish to use object modelling. Having identified critical business factors, the business processes must be mapped and understanding arrived at for key factors such as what data will be available when and in what form? How will the output from the model be used? Who will carry out the modelling and how will the model be maintained over time? By answering these strategic questions, the firm can pick and choose from the software available and develop it to suit its needs. It is important that at all times its objectives in introducing modelling are clear and that as development progresses the model's ability still fulfils these objectives. Without this link into the business process, the model's potential to impact on the business is not maximised. By harnessing knowledge in this way, a firm can strive for excellence in their particular field.

As the use of the technology becomes more widespread and automatic data transfer more efficient, the integration of firms into a cohesive temporary project organisation will take place. This aggregation of forward thinking firms within project teams will deliver excellence within the project environment and hence develop the ability to deliver the innovative, technically complex constructions that the 21st century will demand. The vision that was portrayed in chapter 4 of modelling becoming central to the project process will come closer to realisation, but a key factor will be that individual firms want it that way because it is the most efficient, cost-effective way for their business to operate.

As the application of object modelling within the construction industry grows, a number of key areas will need to be developed to support its use. Most importantly, there are issues of training and education which need to be put in place. The remainder of this chapter considers these and how the implications of adopting object modelling need to be considered.

Training for object oriented modelling

In view of the sophistication of the technology required to use object modelling, it is unlikely that construction firms or professionals will have the level of IT skills necessary to introduce and operate an object modelling system. Equally, IT specialists who are able to operate an object modelling system will not be aware of the special characteristics of construction work let alone the nuances of the construction project life cycle.

In the short term two solutions are possible.

(a) Providing intensive training courses in object modelling for clients, professional designers and construction personnel to familiarise them with the system. As learning about object modelling is an iterative process, continual access to an appropriate computer system is essential. It will be expensive and slow but may ultimately be the cheapest way of introducing a change of attitude within a highly conservative construction industry. The special knowledge base and skills of the professions will enable maximum exploitation of object modelling.

(b) Bringing in IT specialists means that, in theory, the object modelling system can start running to its full potential immediately provided that there are dedicated professional construction people on hand to provide the project information and offer expert advice. This is certainly the quickest way to introduce object modelling but may be expensive because of double manning, and depends upon the quality and degree of communication between the key players; there may be a language problem.

The role of education

The long-term development of object modelling skills must come through the introduction of special modules into construction undergraduate and postgraduate courses at universities and colleges. The next generation of architects, engineers and construction managers must be object modelling literate when they finish their degrees so that they can immediately start to exploit the potential of object modelling in their jobs. Some universities are already including object modelling in the curriculum and will attract the brightest graduates on construction courses traditionally seen as moribund and low tech.

Research is proceeding both on the principles and on the practice of object modelling. There are opportunities for both blue sky research by universities and applied research by industry to push back the frontiers of knowledge in this revolutionary approach to planning, designing and constructing the buildings of tomorrow.

Chapter 7

Who will benefit?

n the past over-eager evangelists have attempted to sell the benefits of IT to a sceptical construction industry. Not only has the capability and reliability of IT been oversold, but also the often difficult technical issues of compatibility of hardware and software, robustness and the complexity of the programming/systems task have been ignored in the enthusiasm for the technology. In addition, predictions about the take up and the effect that the technology will have have often been optimistic. It is undeniable that computers and ever more sophisticated software have made big inroads, in particular, the ever more powerful stand alone PC. However the construction process has so far been surprisingly unaffected by the onslaught of the technology. Largely, it is argued, as a result of the fragmented nature of the industry, no one group has had either the influence or the commercial motivation to apply IT to other than their own specialism. As a result, IT has been applied in a narrow, focused way to traditional methods of doing business. The tendency has mainly been to computerise the traditional processes or ways of working.

This contrasts with less fragmented industries where, either due to major companies assembling under one roof, such as the car industry, or a few giant powerful clients, such as the petrochemical industry, the manufacturing process has been viewed far more holistically and the opportunity to fundamentally rethink the whole or substantial parts of the traditional way of working has been taken. In such instances, new methods that harness the full potential of IT have been developed. It is in instances such as this that major improvements in productivity, quality, cost reduction or time to market have been achieved and measured objectively.

Those major UK construction clients who operate internationally have realised both their power to influence the industry and their ability to compare on a country by country basis the value for money that they receive. The trend towards global operation and country comparisons using ever better benchmarking methods will cause these clients to pressurise the UK construction industry to dramatically improve its performance. Thus, for these clients there will be an expectation of performance improvements in the construction activity they purchase that they have already seen elsewhere in their business. They will not understand why, for example, supply chain management techniques, partnering methods, the use of standard components and a continuous improvement philosophy are not routinely used by their suppliers in construction when they are used in other, similarly complex industries, that they regularly deal with. For clients, the use of electronic prototypes, visualisation and the harnessing of user knowledge will result in more certainty in terms of cost, time and quality targets, and the gradual improvement in value for money, operating costs and buildings that incorporate lessons learned elsewhere. Better preplanning, in all its forms, will slowly achieve the improvements, not from cutting fees, scope and specification, or from robbing players of a reasonable profit, but from performance enhancements that gradually eat into the current built-in inefficiencies that have evolved with the traditional fragmented system.

It is clearly difficult to predict precisely what the benefits to each of the traditional construction industry parties will be. However, educated guesses can be made, in particular when observing the effect on key players in other design-to-order industries. A first benefit, if only to some, will be the *'it's in the mind'* test. Some key players in other industries have not been able to make the mental leap to *'we can and must do it'* but have settled for the familiar and apparently comfortable opinion that their industry is different and that *'it would not work'*. Boeing acknowledged that such attitudes exist even among their own managers and regular framework partners. After extensive trials and consultation, managers at the most senior level drew a very public and symbolic line in the sand and committed their company irrevocably to the new way of working.

For those with the will and perhaps courage to make the new technology work for them, there will be considerable benefits, not least that they are able to obtain for their clients, products or services that represent better value for money and for their own company a clear competitive advantage.

For non-engineering designers such as architects, the real benefit will come from screen based 3D design from the earliest stage. Creative ability will be enhanced as the limitations of pencil and paper design methods are shaken off. The technology will enable the automation of much of the tedious design issues, such as dimensional coordination, component specification and schedules, as Java-based electronic catalogues containing intelligent generic objects are accessed through private pages on the Internet and incorporated seamlessly into the design. This will release designers to do what humans are good at – thinking laterally – and allow them, using the electronic prototypes they produce, to design buildings ever more closely to the proposed users' needs. Indeed, one designer already using the technology stated enthusiastically that

> the model/prototype concept allows me to protect much more effectively those parts of the design that I consider to be important.

For engineering designers, this technology will enable the development of generic engineering systems, such as intelligent generic steel frames, that include beam and column objects which, in effect, design themselves. Again, by harnessing the rules of physics that relate to structural or HVAC design into the machine it will free up designers to develop more innovative or efficient systems.

For contractors, specialist trade contractors and component manufacturers and suppliers, it will be much easier for them to relate to modern advanced manufacturing techniques such as lean manufacturing and mass customisation. The development of digital mock-ups and generic objects such as a lift, an escalator or a cladding panel will allow contractors, specialist trade contractors, component manufacturers and suppliers to undertake better fit testing, assembly trials and planning.

Project managers and those specialising within the discipline in construction management face a big challenge to identify and adapt advanced manufacturing techniques in the area of advanced planning, pre-assembly component fit checking and planning assembly trials with digital mock-up tools. Design management, supply chain and assembly management will also need to be developed using lessons from advanced manufacturing. With the gradual acquisition of these skills and the reduction in the construction industry's management skills cultural lag, project managers and construction managers will be able to achieve the sort of process improvement challenges set by Latham and, in so doing, develop a competitive advantage through operating world class systems and techniques.

Potentially one of the greatest changes brought about by the technology outlined in this report is in the area of cost management. Predictive cost modellers using object oriented technology will base cost estimates, even at a quite early stage, on the likely cost of construction by effectively building within the computer in minutes the proposed building in concept and pricing it by using a maintained cost database and identifying cost drivers and risk and adjusting for them. It will allow elemental analysis as the design of different components takes place. In addition, linked to the design process and the various forms of the project model it will provide designers with a rapid feedback on the cost implications of design changes or options. This will revolutionise the service that a cost manager is able to give the rest of the design team.

Building operators will benefit simply because their buildings will have most of the normal teething problems ironed out as electronic prototypes become used increasingly to greater effect. In addition, benefits for operators will be achieved as a result of their reaction to early designs being taken on board. Their experienced eyes will lead to a gradual but continuous improvement generally in the design and construction process. All of this will mean a change in relationships between parties that will at times be radical. This change will need both suitable training and the management of the change. None of this will come cheap in terms of time, effort and cash. Considerable investment not only in training but also in developing both the technology and the processes to meet company/practice needs will be necessary. This should not be understated. However, experience elsewhere shows that if properly undertaken the competitive advantage achieved makes it all a worthwhile investment.

Lastly, perhaps the greatest benefit to all parties will result from the better coordination and management of project data and the design process itself. This will be achieved by the adoption of concurrent engineering techniques and what have been called in the US petrochemical industry *global 24 hour virtual design offices*. Simply, that is when designers have the ability to log in their work from any geographical location, at any time, as they complete a section of it. This work will be automatically integrated into the master model and, using clash detection routines and concurrent engineering methods, the design manager can look at the implication on the model so far of each new piece of design work and how it relates to the agreed project brief.

Glossary

In this publication we have used a more relaxed style of writing in an attempt to make the publication more readable. Thus we have tried to use a balanced mixture of the more common terms that are associated with object orientation and project modelling. We have resisted the temptation to just use the one term for a particular thing and have used a limited number of the more common to aid wider understanding. We are clear, however, that there will soon be a need, as the level of sophistication and usage increases in the industry, to agree one common term for each situation or technique. We are of the view that, regrettably, that point has not yet been reached.

3D Studio, 3D Studio Max A modelling and rendering system with the ability to produce animation and video.

Algorithm A logical arithmetical or computational procedure that if correctly applied ensures the solution of a problem. In the context of this publication it is used to describe design or facilities planning rules that have been codified elsewhere.

Applet A self-contained reusable computer application.

ASCII American Standard Code for Information Exchange: computer code for representing alphanumeric characters.

Attribute A fundamental concept which is used in most data models. The attributes of an object are defined in a class and values are given separately for each instance. In many data models the data types of attributes can be either basic (integer, real, string) or complex.

AutoCAD A leading and industry standard in CADD from AutoDesk Ltd.

Bar-coding A method of graphically representing computer binary code using a series of alternating dark and light bars. This sequence of bars can be read by hand-held or flat-bed scanners to enable rapid data entry without the need for keyboards.

CAD Computer aided design.

CADD Computer aided drafting and design.

CAE Computer aided engineering.

CAM Computer aided manufacturing.

Concurrent engineering A method of working which allows two or more processes to be carried out in parallel.

CYMAP Building services software.

Design-to-manufacture Where design is primarily for batch production, whether small, medium or large scale.

Design-to-order Where each design is for a 'one-off' project. Often the uniqueness of the project is in the resulting solution rather than in the components used, design algorithms or skills employed to design and build.

Digital mock-up, DMU A computer-based version of the traditional mock-up or model. This allows the realistic testing of assembly methods and components on a computer.

DXF Data exchange format used in AutoCAD. A file format that is mainly in use to import and export CADD data between AutoCAD and other CADD related programs.

Electronic objects Used to differentiate from 'real world' objects.

Electronic prototype A prototype assembled in computer space from electronic objects.

ICAD A software development system by Concentra Ltd which allows the development and assembly of knowledge based objects using Generative Technology.

INTEGER Ltd A computer software company, originally started by Ove Arup employees.

ISDN Integrated Services Data Network: allows the sending and receiving of high volumes of assorted data, text, voice and image, through a single digital telephone line over the public telephone network.

Java An open, standard, universal platform for network computing that scales from telephones to major mission critical applications, building on existing hardware and software architectures.

KBE Knowledge based engineering: ICAD is an example of a KBE system. It usually refers to a software 'platform' that is widely available and used, in contrast to the writing of a programme from first principles. It is, in principle, similar to buying an 'off the shelf' tested spreadsheet package, rather than writing a personal spreadsheet package from scratch.

MICRODRAIN Drainage design software.

MicroStation A leading CAD system, owned by Bentley Systems Ltd and widely distributed by Intergraph Ltd.

ModelView A model image rendering system from Intergraph Ltd.

Object Something you can do things to. An object has state, behaviour and identity; the structure and behaviour of similar objects are defined in their common class. The terms 'instance' and 'object' are interchangeable.

Object orientation, OO The notion of describing the world using objects. The term is used in modelling and software development.

Object oriented model A computer model constructed using object oriented methods.

PowerProject Project planning software.

Process A mechanism which transforms an input to an output using identified resources and upon which some control may be exerted.

Process improvement The improvement of a current business/design process, often by identifying a new way of working that usually with the use of advanced IT will reduce the time/man-hours required to undertake a task. Important benefits of such a process are: cost reduction, better coordination of data, better quality control procedures and shorter delivery times. Process improvement targets are usually benchmarked as 'industry best practice' or as 'world class'.

Process mapping An analysis tool for structuring process information to provide an holistic view of a process.

Process reengineering The ability to change the sequence and substance of processes in order to improve at a strategic level.

Project modelling Object modelling applied to a project and including more information than just geometry.

Project process A way of viewing a project life cycle using systems theory.

Reflex An object oriented modelling system, now owned by Parametric Technologies Inc., which can be driven like a very user-friendly CAD system. Now reissued as ProReflex.

Smart dot technology The building of a powerful microchip onto a small dot, usually smaller than the head of a traditional drawing pin. This allows data to be stored, accessed and downloaded on to such a dot. Usually, the dot is attached to a paper or plastic strip with adhesive on one side, which allows it to be stuck to, for example, a machine component, an airline ticket or passenger luggage.

Sonata An object based CAD system developed from the original RUCAPS system.

STEP Standard for the Exchange of Product Model Data (ISO 10303).

SuperScape A tool for producing virtual reality models.

UNIX A computer operating system based on open standards, and able to be specially compiled and adapted to run on various computer platforms.

Virtual Something that is 'virtual' is something that appears to be real (or at least is given the intention to seem real), but is just an illusion of its real counterpart.

Virtual reality An attempt to reproduce real life sensations using a digital medium, allowing the user to experience projected or imagined real world experiences. Most forms of virtual reality rely on visual, motile and tactile feedback in response to a user's input. Visual feedback can be via conventional computer screen or head mounted optics.

Virtual design office With the use of advanced design technology, it is possible to link designers from around the world, each working independently on a part of a design, to one project master model located in a client's office or the office of the project manager. Thus time (zones) and geographical location begin to become less important.

VRML Virtual Reality Modelling Language: A file format for describing interactive 3D objects and worlds to be experienced on the World Wide Web. The first release of the VRML 1.0 Specification was created by Silicon Graphics, Inc. and based on the Open Inventor file format. The second release of VRML adds significantly more interactive capabilities.

WWW World Wide Web: One of the most popular ways of accessing the Internet, which allows users to browse (or 'surf') through information using hypermedia links through its text and image.

References and further reading

References

The following publications have directly influenced the authors' thinking and support the contents of the text.

1 WS Atkins (1994). *Construction Strategies*. A report to the European Commission.

2 There are several versions of the history of the Internet, we have followed that in *The Economist*. An alternative version can be found in *New Scientist*.
Anon (1996). A World Gone Soft - A Survey of the Software Industry. *The Economist*, 25 May.
Anon (1994). What a Tangled Web They Wove. *New Scientist*, 30 July.

3 Christiansen T. and Levitt R. (1992). *The Virtual Design Team*. CIFE, Stanford University, p46.

4 Levitt R. E., *et al*. (1994). *The Virtual Design Team: Using Computers to Model Information Processing and Communication in Organisations*. CIFE, Stanford University, p19.

5 Levitt R. E., *et al*. (1994). *The Virtual Design Team: A Computational Simulation Model of Project Organisations*. CIFE, Stanford University, p35.

6 Yan Jin and Levitt R. E. (1996). *The Virtual Design Team: A Computational Model of Project Organisations*. CIFE, Stanford University, p32.

7 Checkland P. B. (1981). *Systems Thinking Systems Practice*. Wiley, Chichester.

8 Bennett R. J. and Chorley R. (1978). *Environmental Systems*. Methuen, London.

9 Fisher G. N. and Chen L.Y. (1992). *Information Management in a Contractor*. Thomas Telford, London.

10 BAA plc (1995). *The Project Handbook*. GTS Gatwick Airport.
This is an internal publication showing the company's revised project process.

11 Womack J. P., Jones D. T. and Roos D. (1990). *The Machine That Changed The World.* Rawson Associates/MacMillan, New York.

12 Pine J. B. (1993). *Mass Customization: The New Frontier in Business Competition.* Harvard Business School Press.

13 Papadakis P. (1991). *A Decade of Architectural Design.* Academy Editions, London.

14 Barlow R. P. G., Fisher G. N. and Papantonopoulos S. N. (1995). *The Application of a Knowledge-Based Engineering System to the Planning of Airport Facilities.* Digest No: 1995/129, Institution of Electrical Engineers, 1/1–4.

15 Garnett N., Barlow R. P. G., Papantonopoulos S. N. and Fisher G. N. (1995). The Application of Knowledge Based Electronic Prototyping to a Project Process at Concept Phase, Modelling of Buildings Though Their Life Cycle. *CIB W78-95, Proceedings Publication 180*, Stanford, 415–425.

16 Papantonopoulos S. N., Barlow R. P. G. and Fisher G. N. (1994). Concurrent Knowledge Based Engineering: Strategic Issues in Product Modelling for the Construction Industry. *ECPPM 1994 - Product and Process Modelling in the Building Industry*, Dresden, 91–94.

17 Papantonopoulos S. N., Barlow R. P. G. and Fisher G. N. (1994). The Application of an Intelligent Knowledge Based Engineering System to the Optimisation of Design, Cost and Quality of Finished Products in Construction. Fourth International Conference on: *Factory 2000 - Advanced Factory Automation.* Institution of Electrical Engineers Publication No 398, York, 369–378.

18 Fisher N. (1993). *Construction as a Manufacturing Process?* An Inaugural Lecture. The University of Reading, 35pp.

19 Fisher N. (1992). The Use of Structured Data Analysis as a Design Tool for Computer Integrated Construction. In: Wagter H. (Ed.). *Computer Integrated Construction.* Elsevier, Amsterdam, 181–188.

20 Fisher N. (1991). A Structured Analysis of Data Flow Systems for Client-based Construction Project Management and for Contracting Companies. In: Maxwell J. W. S. (Ed.) *Applications of Information Technology in Construction.* Thomas Telford, London, 63–76.

21 Barlow R. P. G. and Amirudin R. (1996). Structural Steelwork Planning and Design Evaluation - A Knowledge Based Approach. *Construction on the Information Highway - Proceeding Publication 198*, Bled, Slovenia, 69–76.

22 International Alliance of Interoperability (IAI). Information can be found on the Internet at http:www.interoperability.com

23 *Islands of Automation*. Information can be found on the Internet at http://www.vtt.fi/cic/ratas/islands.html

Further reading

This list of publications is intended to allow readers to explore deeper into the subject, and also into a number of related areas.

Anderson A.E. Jr., Heinze W. J. (1992). *C++ Programming And Fundamental Concepts*. Englewood Cliffs, N J, Prentice Hall, ISBN/ISSN/ISSN/0131182668.

Anon. (1996). Near Virtual Reality Creates Walk-Through Environment. *Mer-Marine Engineers Review,* **34**, ISSN 0047-5955.

Belhadj T. A. (1989). *Computer-Aided Architectural Evaluation and Design*. PhD Thesis, University of Sheffield. Ref: 40-4850.

Bernstein D. S. (1995). Secure the Virtual Office. *Datamation*, **41** (1) 49–51, ISSN 0011-6963.

Besant C. B. (1983). *Computer-Aided Design And Manufacture*. 2nd edition, Chichester: Horwood, ISBN 0853125228.

Besant C. B. and Lui C. W. K.(1986). Computer-Aided Design and Manufacture. *Ellis Horwood Series in Engineering Science*, Chichester: Horwood.

Blount G. N., Kneebone S. and Kingston M. R. (1995). Selection of Knowledge-Based Engineering Design Applications. *Journal of Engineering Design*, **6** (1) 31–38, ISSN 0954-4828.

British National Conference on Databases (1989). *Proceedings of the Seventh British National Conference on Databases (BNCOD 7)*: Heriot-Watt University, 12–14, British Computer Society workshop series, Heriot-Watt University, ISBN 0521382815.

Brown A. (1994). The First Annual Conference and Exhibition on Business Process Reengineering. *Journal of Information Technology*, **9** (4) 297–302, ISSN 0268-3962.

Brown H. T. and Evely M. R. (1996). Developing and Improving the New Product Process at Marconi. *Journal of Research*, **13** (2) 91–99, ISSN 0264-9187.

Case K. and Newman S. T. (Eds)(1994). *Tenth National Conference on Manufacturing Research.* 5–7 September 1994. Loughborough University of Technology. Organised under the auspices of COMED, London. Taylor & Francis, ISBN 0748402543.

Coates E. R., Sarker B. R. and Ray T. G. (1996). Manufacturing Set-up Cost Reduction: Nineteenth International Conference on Computers and Industrial Engineering, Miami, FL, 4 - 6. *Computers & Industrial Engineering*, **31** (1–2) 111–114, ISSN 0360-835.

Coyne R. F. (1991). *ABLOOS, An Evolving Hierarchical Design Framework.* PhD Thesis, Carnegie Mellon University, USA.

Dillon T. S. and Tan P. L. (1993). *Object-Oriented Conceptual Modelling.* Prentice Hall, Australia, ISBN 0-13-712952-1.

Duffy F. and Hannay P. (Eds) (1992). *Changing Workplace.* London: Phaidon Press, ISBN 185454120X.

Dwivedi S. N., Verma A. K. and Sneckenberger J. E. (Eds)(1991). CAD/CAM Robotics and Factories of the Future '90, Concurrent Engineering. International Society for Productivity Enhancement. *Proceedings of the Fifth International Conference on CAD/CAM, Robotics, and Factories of the Future (CARS and FOF '90).* Springer-Verlag, USA, 1. 3-54053399-0.

Effective CADCAM. Papers from a conference held at Pembroke College, Cambridge, 29-30 June 1983: Sponsored by the Engineering Manufacturing Industries Division of the Institution of Mechanical Engineers, co-sponsored by the Institution of Production Engineers in association with the CAD Centre, London. Mechanical Engineering Publications, ISBN 0852985177.

ESCAPE-4. Fourth European Symposium on Computer Aided Process Engineering, 1994, University College, Dublin. Institution of Chemical Engineers Symposium Series. Institution of Chemical Engineers.

Fenves S., Flemming U., Hendrickson C., Maher M. L., Quaderl R., Terk M. and Woodbury R. (1994). *Concurrent Computer Integrated Building Design.* Prentice-Hall, Inc., USA, ISBN 0-13-175753-9.

Fritz B. (1991). *Creating Products in the Next Generation Design and Manufacturing Environment.* The Christopher Hinton Lecture, London. Fellowship of Engineering.

Georgakopoulos D., Hornick M., and Sheth A. (1995). An Overview of Workflow Management – From Process Modelling to Workflow Automation Infrastructure. *Distributed and Parallel Databases*, **3** (2) 119–153, ISSN 0926-8782.

Gero J. S. (Ed)(1985). Knowledge engineering in computer-aided design. *Proceedings of the IFIP WG 5.2, Conference on Knowledge Engineering in Computer-Aided Design*, Budapest, Hungary. North-Holland, Netherlands, ISBN 0-444-87825-4.

Gero J. S. (Ed)(1991). *Artificial Intelligence in Design '91*. Butterworth Heinemann, ISBN 0-7506-1188.

Gero J. S. and Sudweeks F. (1996). *Advances in Formal Design Methods for CAD*. Chapman & Hall. ISBN 0 412 72710 2.

Gertsbakh I. B. (1977). Models of Preventive Maintenance. *Studies in Mathematical and Managerial Economics*. Amsterdam; Oxford: North-Holland Publishing Co., ISBN 0720404657.

Graham I. (1994). *Object oriented methods*. 2nd edition, Wokingham: Addison-Wesley, ISBN 0201593718.

Gupta R. and Horowitz E. (Eds)(1991). Object-Oriented Databases with Applications to CASE, Networks, and VLSI CAD. *Prentice Hall Series in Data and Knowledge Based Systems*, Prentice Hall, USA, ISBN 0-13-629833-8.

Haase B. (1994). Neon - A Shining Example of Design Efficiency - with the Help of 3D CAD Models and Concurrent Engineering, Chrysler Speeds its New Car to Market in Record Time. *Computer Graphics*, February, 41–44.

Hahm J. and Lee M. W. (1994). A Systematic Approach To Business Process Reengineering. *Computers & Industrial Engineering*, **27** (1–4) 327–330, ISSN0360-8352.

Henderson-Sellers B. (1992). *A Book of Object-Oriented Knowledge: Object-Oriented Analysis, Design, and Implementation; A New Approach to Software Engineering*. New York; London: Prentice Hall, ISBN 0130594458.

Ho, C. F. (1996). Information Technology Implementation Strategies for Manufacturing. *International Journal of Operations & Production Management*, **16**.

Hoggan P. and Masini G. (Trans.)(1991). *Object-Oriented Languages*. Academic, ISBN 0124773907.

Ince D. (1991). *Object-Oriented Software Engineering With C++*. London: McGraw-Hill, ISBN 0077074025.

Ingham P. C. (1989). *CAD Systems in Mechanical and Production Engineering*. Newnes, ISBN 0434908703.

Jarzabek S. and Ling T. W. (1996). Model-Based Support for Business Reengineering. *Information and Software Technology*, **38** (5) 355–374, ISSN 0950-5849.

Kalpakjian, S. (1992). *Manufacturing Engineering and Technology*. Reading, Mass: Addison-Wesley Pub. Co., ISBN 0201569507.

Kopacek P. E. (Ed.)(1994). Intelligent Manufacturing Systems 1994: *IMS'94*: IFAC Workshop, Vienna, Austria, 13–15 June 1994. 1st Edition, Oxford ; New York: published for the International Federation of Automatic Control by Pergamon, ISBN 0080422357.

Lawless J. A. and Miller M. M. (1991). *Understanding CLOS, The Common Lisp Object System*. Digital Equipment Corporation, USA, ISBN 1-55558-064-5.

Lawson A. and Horton C. (1990). *Manufacturing Technology*. Delmar Technology Series, Albany, N Y: Delmar Publishers, ISBN 082733462.

Lawson M. and Karandikar H. M. (1994). A Survey of Concurrent Engineering. *Concurrent Engineering - Research and Applications*, **2** (1) 1–6, ISSN 1063-293.

Lei D., Hitt M. A. and Goldhar J. D. (1996). Strategic Flexibility. *Organisation Studies*, **17** (3) 501–523, ISSN 0170-8406.

Leinhardt S. (1996). Information Technology and Business Process Reengineering. *International Journal of Micrographics & Optical Technology*, **14** (1) 5–6, ISSN 0958-9961.

Li H. (1996). The Role of IT Manager in Construction Process Reengineering. *Building Research and Information*, **24** (2) 124–127, ISSN/0961-3218.

Luyben W. L. (1990). *Process Modelling, Simulation, and Control for Chemical Engineers*. McGraw-Hill Chemical Engineering Series, McGraw-Hill, ISBN 007039159.

Maba M. L., Falkenburg D. R. and Hill D. H. (1996). Technology Management and American Culture – Implications for Business Process redesign. Wayne State University, Dept Anthropol, Detroit, MI, 48202. *Research-Technology Management*, **39** (6) 44–54, ISSN 0895-6308.

Magnusson B., Meyer B. and Perrot J.-F.(Eds). TOOLS Europe '93. Versailles, France. Proceedings of the Tenth international conference. *Technology of object-oriented languages and systems*. New York; London: Prentice Hall, ISBN 0130971146.

Maher M. L. and Rutherford J. (1993). Video Conferencing and CAD, Architectural Science: Past, Present and Future. *ANZAScA*, in Hayman S. (Ed.), University of Sydney, Australia., 113–118.

Mahoney D. P. (1996). Virtual-Reality on the PC. *Computer Graphics World*, **19** (5) 53, ISSN 0271-415.

Mahoney D. P. (1996). Reality Meets Virtual-Reality. *Computer Graphics World*, **19** (7) 18, ISSN 0271-415.

Makino H. (1983). Assembly Automation: *Proceedings of the Fourth International Conference*, October 11–13, 1983, Tokyo, Japan: Japan Society of Precision Engineers, Bedford: IFS (Publications); Oxford: North-Holland, Publishing, ISBN 0903608480.

Masini G. et al. (1991). Object-Oriented Languages. Translated by Hoggan P. and Masini G., *The A.P.I.C.*, **34**, Academic, ISSN 0124773907.

McGoldrick P. F. (Ed.)(1986). Advances in Manufacturing Technology: *Proceedings of the First National Conference on Production Research*, University of Nottingham, September 1985, National Conference on Production Research, COPED, IProdE, London: Kogan, ISBN 1850910391.

Mills R. (1994). 3D Modelling – New-Look CAID. *Computer-Aided Engineering*, **13** (8) 12, ISSN 0733-353.

Nadiadi Y. (1990). *An Automatic Floorplanning System for Use in the Interactive Architecture*. PhD Thesis, Essex University, Ref.: 40-2134.

Nevins J. L.,. Whitney D. E. and De Fazio T. L. (Eds)(1989). *Concurrent Design of Products and Processes: a Strategy for the Next Generation in Manufacturing*. New York; London: McGraw-Hill, ISBN 0070463417.

Olthoff W. (Ed.) (1995). *ECOOP '95, Lecture Notes in Computer Science*; 952, Berlin; New York: Springer, ISBN 3540601600.

Ramani K. V., Yap R. and Pavri F. (1995). Information Technology Enables Business Process Reengineering. *Journal of Strategic Information Systems*, **4** (1) 81–88, ISSN 0963-8687.

Rogers P. (1989). *Object-Oriented Modelling for Flexible Manufacturing Cells*. PhD Thesis, Cambridge University, Ref.: 39-4733.

Sarkozy F. (1994). The GIS Concept and 3-Dimensional Modelling. *Computers Environment and Urban Systems*, **18** (2) 111–121, ISSN 0198-971.

Sata T. (Ed.)(1989). *Organisation of Engineering Knowledge for Product Modelling in Computer Integrated Manufacturing*. The Second Toyota – A Collection of Contributions Based on Lectures Presented at the 2nd Toyota Conference, Aichi, Japan, 2–5 October 1988, Elsevier, USA, ISBN 0-444-88149-2.

Scherer R. J. (Ed.)(1994). Product modelling in the building industry. *Proceedings of ECPPM '94 - The first European conference on product and process modelling in the building industry* - Dresden, Germany, 5–7 October, ECPPM' 94. A A Balkema, Rotterdam, Netherlands. ISBN 90 5410 584 8.

Sodhi R. S., Zhou and Das S. (Eds)(1994). Advances in Manufacturing Systems: Design, Modelling, and Analysis, *9th International Conference on CAD/CAM, Robotics, and Factories of the Future (CARS & FOFS 93)* held in Newark, New Jersey, USA on August 17-20, 1993, Manufacturing Research and Technology, Amsterdam ; New York: Elsevier, 22.

Teicholz P. and Fischer M. (1994). Strategy for Computer Integrated Construction Technology. *Journal of Construction Engineering and Management*, **120**,117–131.

Topping B. H. V. and Papadrakakis M. (Eds)(1994). Artificial intelligence and object oriented approaches for structural engineering. Some of the papers presented at the *Second International Conference on Computational Structures Technology*, Athens 30 August to 1 September, Edinburgh: Civil-Comp, ISBN 0948749296.

Tsichritzis D. C. (Ed.) (1985). Office Automation: Concepts and Tools, *Topics in Information Systems*. Berlin: Springer, ISBN 354015129X.

Tzafestas S. G. (Ed.)(1989). Knowledge-Based System Diagnosis, Supervision, and Control. *Applied Information Technology*, New York; London: Plenum.

Uhlig R P, Farber D. J. and Bair J. H. (1981). The Office of the Future: *Communication and Computers*, Monograph series, Amsterdam; Oxford: North-Holland.

Voss C. A.(Ed.) (1987). Just-in-time Manufacture. *International Trends in Manufacturing Technology*, IFS, ISSN 0948507497.

Wang W. (1989). *A Knowledge-Based Modelling System for the Design and Evaluation*. PhD Thesis, Loughborough University, Ref: 39-2125.

Wildemann H. (1995). Integrated Information-Systems - A Critical Success Factor in Customer-Focused, Lean Companies, Germany. *Wirtschaftsinformatik*, **37** (2) 95–104, ISSN 0937-6429.

Wu B. (1992). *Manufacturing systems design and analysis*. London: Chapman & Hall, ISBN 0412408406.

Yau, Nie-Jia (1992). *An Object-Oriented Project Model for Integrating Building Design, Construction Scheduling, and Cost Estimating for Mid-Rise Construction*. University of Illinois at Urbana-Champaign (0090).

Yu Esk, Mylopoulos J.and Lesperance Y. (1996). AI Models for Business Process Reengineering, Canada, IEEE. *Expert-Intelligent Systems and their Applications*, **11** (4)16–23, ISSN 0885-9000.

ACT-Reading

Advanced Construction Technology - Reading (ACT-Reading) is a focused multi-disciplinary research group within the Department of Construction Management and Engineering at The University of Reading. It has several broad areas of research expertise: Advanced IT and systems; process mapping/reengineering; benchmarking; customer/user focus in early design; and facilities management issues. The department of which is a part recently achieved a 5* score, the top rating, in the last research assessment exercise. This is in effect an independent benchmark of 'world class' research performance.

Under the direction of Professor Norman Fisher, Richard Barlow and Naomi Garnett, ACT-Reading has a major research programme working with major clients, designers, contractors, other research organisations, major industrial groups and component manufacturers. All current research projects are funded between 50–100% by industry. ACT-Reading is based in its own specially refurbished premises and has the research theme of 'construction as a manufacturing process'. A research community has been carefully created that allows academic and industrial based researchers to work as a team on important research issues and at the highest levels of scholarship. The work of ACT-Reading can be viewed on the Internet at: http://www.act.rdg.ac.uk.

Taylor Woodrow Management Ltd

Taylor Woodrow Management Ltd, part of the Taylor Woodrow Group of Companies, is a leading management and consultancy group committed to managing projects in new and innovative ways in order to provide added value to its customers.

Object modelling is seen as a key management tool for providing not only the visualisation of the concept but also the basis to control the development and implementation of a construction project. Taylor Woodrow is addressing some of the key issues that will be facing the industry in the 21st Century. The development of a prototype and data exchange with the supply chain is seen as the next major criteria. Taylor Woodrow is the biggest spender on research and development in the building and construction industry. Taylor Woodrow Management is committed to the provision of world class services. This study is but one example of Taylor Woodrow's commitment to the industry, its customers and suppliers. Web site: http://www.Taywood.co.uk